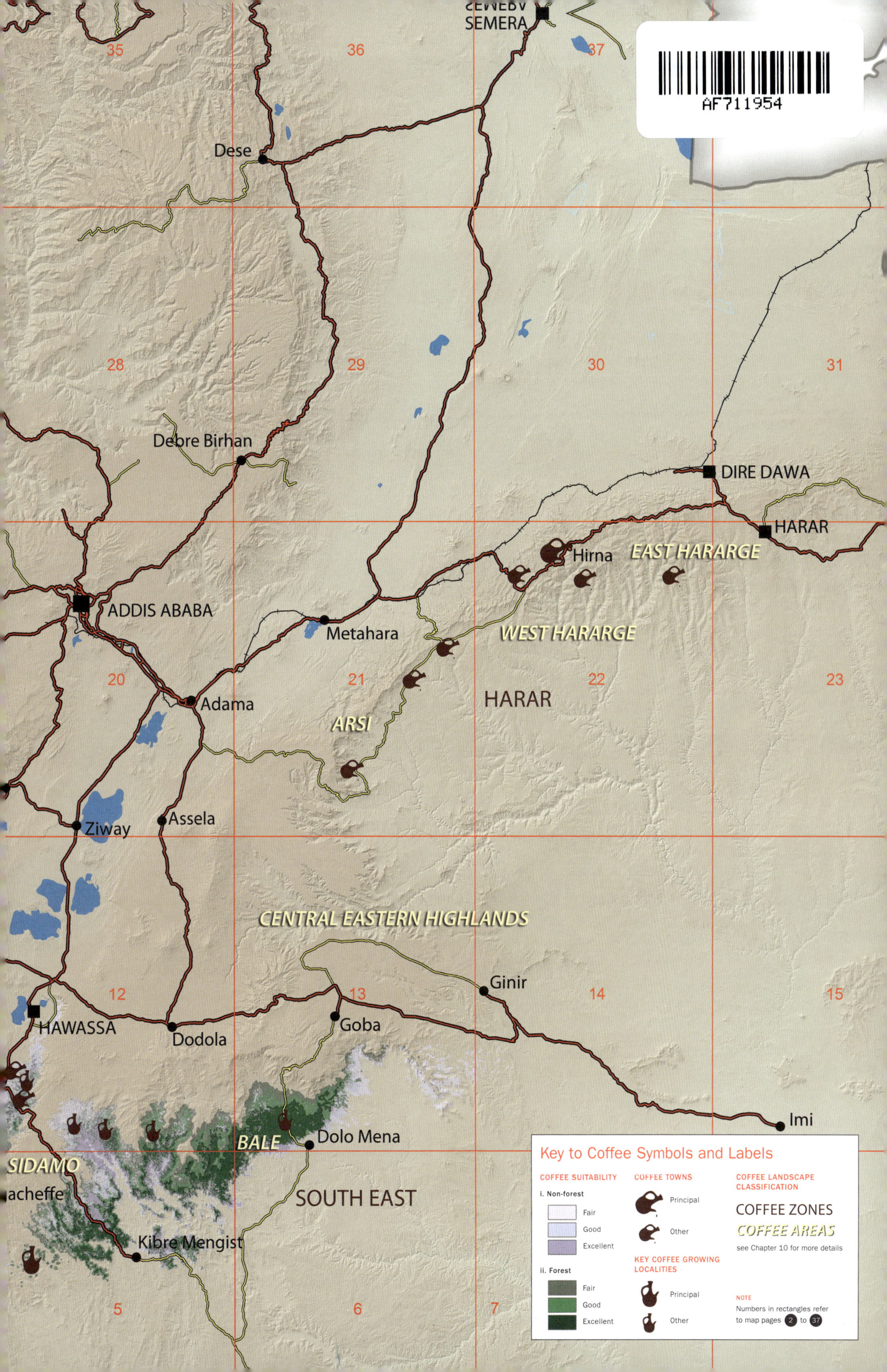

SEMERA
Dese
Debre Birhan
ADDIS ABABA
Metahara
Adama
Assela
Ziway
DIRE DAWA
HARAR
Hirna
EAST HARARGE
WEST HARARGE
HARAR
ARSI
CENTRAL EASTERN HIGHLANDS
Ginir
HAWASSA
Dodola
Goba
Imi
BALE
Dolo Mena
SIDAMO
acheffe
SOUTH EAST
Kibre Mengist
35
36
37
28
29
30
31
20
21
22
23
12
13
14
15
5
6
7
Key to Coffee Symbols and Labels
COFFEE SUITABILITY
i. Non-forest
Fair
Good
Excellent
ii. Forest
Fair
Good
Excellent
COFFEE TOWNS
Principal
Other
KEY COFFEE GROWING LOCALITIES
Principal
Other
COFFEE LANDSCAPE CLASSIFICATION
COFFEE ZONES
COFFEE AREAS
see Chapter 10 for more details
NOTE
Numbers in rectangles refer to map pages 2 to 37

Coffee Atlas of Ethiopia

Aaron P. Davis, Tim Wilkinson,
Zeleke Kebebew Challa,
Jenny Williams, Susana Baena,
Tadesse Woldemariam Gole
and Justin Moat

Kew Publishing
Royal Botanic Gardens, Kew

Royal Botanic Gardens Kew

First published in 2018 by Royal Botanic Gardens, Kew
Reprinted 2018, 2020
Royal Botanic Gardens, Kew, Richmond, Surrey, TW9 3AB, UK
www.kew.org

Distributed on behalf of the Royal Botanic Gardens, Kew in North America by the University of Chicago Press, 1427 East 60th Street, Chicago, IL 60637, USA

ISBN 978-1-84246-660-5
eISBN 978-1-84246-661-2

British Library Cataloguing in Publication Data
A catalogue record for this book is available from the British Library.

Copy-editing: Sharon Whitehead
Design and page layout: Christine Beard
Production management: Andrew Illes, Georgie Hills

Front cover: Harvested coffee fruits (cherry), Illubabor, south-west Ethiopia.
Image: Alan Schaller (courtesy of Union Hand-Roasted Coffee)
Back cover: (from top to bottom) Coffee leaf tea, coffee husks and dried coffee fruits; photograph taken in the city of Harar. Image: Jenny Williams

Printed and bound in the UK by Gomer Press Limited

FSC www.fsc.org MIX Paper from responsible sources FSC® C114687

The paper used for this publication is FSC certified

Funded by the SCIP Fund

UKaid from the British people

NORWEGIAN MINISTRY OF FOREIGN AFFAIRS

MINISTRY OF FOREIGN AFFAIRS OF DENMARK

Proceeds from the sale of this book go towards the development of the *Coffee Atlas of Ethiopia*.

For information or to purchase all Kew titles please visit shop.kew.org/kewbooksonline or email publishing@kew.org

Kew's mission is to be the global resource in plant and fungal knowledge and the world's leading botanic garden.

Kew receives about one third of its running costs from Government through the Department for Environment, Food and Rural Affairs (Defra). All other funding needed to support Kew's vital work comes from members, foundations, donors and commercial activities, including book sales.

Contents

Contributors, partners and acknowledgements

Authors

The authors are Aaron Davis, Tim Wilkinson, Justin Moat, Jenny Williams and Susana Baena (Royal Botanic Gardens, Kew, UK), and Zeleke Kebebew Challa and Tadesse Woldemariam Gole (Environment and Coffee Forest Forum (ECFF), Addis Ababa, Ethiopia). The bulk of the cartography (Atlas map production) was undertaken by Tim Wilkinson. This work should be cited as: Davis, A.P., Wilkinson T., Challa, Z.K., Williams, J., Baena, S., Gole, T.W. & Moat, J. (2018). *Coffee Atlas of Ethiopia*. Royal Botanic Gardens, Kew (UK).

The SCIP Fund

The *Coffee Atlas of Ethiopia* is a product of the project *Building a Climate Resilient Coffee Economy for Ethiopia*, conducted under the Strategic Climate Institutions Programme (SCIP) Fund. The SCIP Fund was designed to build Ethiopia's capacity to cope with climate change across the public, private and civil society sectors and to respond to the challenges of transitioning to a Climate Resilient Green Economy (CRGE). The SCIP Fund was financed by the Governments of the UK, Norway and Denmark.

Acknowledgements

We are grateful to in-country project partners: Ethiopian Biodiversity Institute (EBI), National Meteorology Agency (NMA), Ministry of Environment Forest, and Climate Change (MEFCC), Ministry of Agriculture (MoA), Addis Ababa University (AAU) and the Oromia Coffee Farmers' Cooperative Union (OCFCU). In Ethiopia, we thank the following organisations and individuals: KPMG (Ethiopia), the Department for International Development (DFiD) and the Ethiopian Commodity Exchange (ECX); Dagne Chomen (OCFCU), Sebsebe Demissew (AUU), Paul Stewart (TechnoServe), Ben Irwin (KPMG), Yvonne and Guy Levene (Bale Mountain Lodge), Adam Overton and partners (Gesha Village Coffee Estate), and the proprietors of Aregash Lodge (Yirga Alem) and Selam Pension (Yayu). At the Royal Botanic Gardens, Kew, we thank Robert O'Sullivan (for fieldwork support), Amanda Cooper (for providing assistance with image processing in ENVI), and Russell Smith (for finance support).

We would like to thank the following people for undertaking sensory evaluation of Ethiopian coffees: Jeremy Torz and Steven Macatonia (and team) at Union Hand-Roasted Coffee (London); Anette Moldvaer at Square Mile Coffee Roasters (London); Henry Clifford at W.R. Wakefield (London). Thanks also to Stephen Leighton at Has Bean Coffee (Stafford), and Dagne Chomen (OCFCU, Ethiopia), for supplying us with coffee samples, tasting notes, and coffee farming locality information.

We are also grateful to: Alan Schaller and Emily Garthwaite, for allowing us to use their stunning photographs (courtesy of Union Hand-Roasted Coffee); Steven Macatonia and Jeremy Torz for reviewing the Atlas (Union Hand-Roasted Coffee); Emma Sage (Speciality Coffee Association (SCA)); David Gower for providing additional road mapping data; and Ib Friis, Sebsebe Demissew, Paulo van Breugel for valuable advice, and for allowing us to reproduce their vegetation mapping data for Figure 5 (page 28).

We are sincerely grateful to the coffee farmers and coffee farming communities across Ethiopia for their generous hospitality and assistance during our fieldwork.

Tigray (northern Ethiopia)

Image: Jenny Williams

1. Introduction

When we talk about Ethiopia and coffee, we are referring to one species only: Arabica coffee (*Coffea arabica*). In its wild state Arabica coffee is a forest plant restricted to the highlands of Ethiopia and neighbouring South Sudan[1]. It has been used in Ethiopia as a food and beverage for many hundreds, if not thousands, of years. Early use was no doubt restricted to gathering fruits from wild forests, with semi-domestication probably coming soon afterwards. Today, Arabica coffee provides a quarter of Ethiopia's export earnings[2] and livelihoods for around 15 million coffee farmers[2–4]. For many of these farmers, coffee is their single most important source of income. Around 50% of the coffee produced within Ethiopia stays within the country, for consumption by Ethiopians. Indeed, coffee drinking is a part of everyday life in Ethiopia, and a key component of Ethiopian culture. As well as the well-known Ethiopian coffee ceremony, coffee is used at major events such as marriages and births, at regionally specific celebrations, and as a medium to build and sustain relationships between family, friends and the community. In Ethiopia coffee is almost universally known as *buna* or *bunna*.

Ethiopian coffee is world-renowned for its broad range of diverse and unique flavour profiles (smell and taste). There are two main reasons for this. First, Ethiopia is the world's main storehouse of genetic diversity for Arabica coffee, and this diversity is manifest in the chemicals that produce flavour. Second, Ethiopia has a complex climate and landscape, resulting in regional and local differences across the coffee-growing landscape. The geographical locations associated with the various flavour profiles are often referred to as 'origins', a term used similarly to *terroir* in the wine industry. Strictly speaking, terroir refers to environmental features of a region or location and their influence on taste and other sensory characteristics. Included amongst the well-known coffee origins of Ethiopia are Harar, Limu, Nekemte, Sidamo and Yirgacheffe, but there are numerous lesser-known origins, which are also unique and distinct. The wide spectrum of flavour profiles adds a unique element to Ethiopian coffee, and makes it especially well-suited to the demands of the high-quality 'speciality' coffee sector. The vast majority of Ethiopian coffee is organic, by default, as most production systems use no synthetic fertilisers, herbicides or pesticides. In addition, because of the intimate association with natural forests, Ethiopian coffee usually has substantial environmental benefits, as outlined below.

Ethiopian coffee is mainly grown under the shade of trees (shade or forest coffee), either within forest or forest-like environments or in farming systems that incorporate specific shade plants, usually indigenous (native) trees or sometimes fruit trees and other crop plants.

Forest is important for coffee because it provides the right conditions for healthy plant growth and successful cropping. It also provides key environmental services, including nutrient recycling, soil preservation, watershed protection, pollination services, temperature buffering, shelter from wind and heavy rainfall, and carbon storage. Ethiopian coffee forests also provide a home for a diverse assemblage of plants and animals, and are thus important for biodiversity preservation.

It is evident, then, that Ethiopian coffee production has many positive attributes, for the consumer and the environment.

Overview

In the *Coffee Atlas of Ethiopia*, our aim is to provide a detailed mapping resource for both farmed and wild Arabica coffee. The Atlas text provides details of the methods used for the production of the maps, as well as background information on geography, the use and consumption of coffee, the botany of the Arabica coffee plant, the coffee-growing climate and agroecology, coffee farming and processing, the coffee-growing areas, and scope and use of the Atlas. At the end of the Atlas there is a gazetteer (geographical name index), to help the user find the geo-political zones, features of the coffee landscape (e.g. key coffee origins, coffee towns and coffee delivery centres), regional capitals, airports, protected areas (e.g. wildlife reserves), major lakes, summits, and towns (settlements).

The maps in the *Coffee Atlas of Ethiopia* show where Arabica coffee is farmed, and could be farmed, and where it is found naturally (i.e. in its wild state). They also show the places where shade coffee is produced, in forest and forest-like environments (such as forest coffee and agroforestry). Coffee garden areas, where either little or no shade is employed, are depicted with a coffee bean symbol. Humid forest (mostly evergreen and rainforest-like) but with no present-day suitability for coffee is also mapped in the Atlas.

LEFT
Drying washed (parchment) coffee on a raised bed (Yayu, south-west Ethiopia)
Image: Alan Schaller (courtesy of Union Coffee)

RIGHT
Travelling across the drylands of the Rift Valley
Image: Jenny Williams

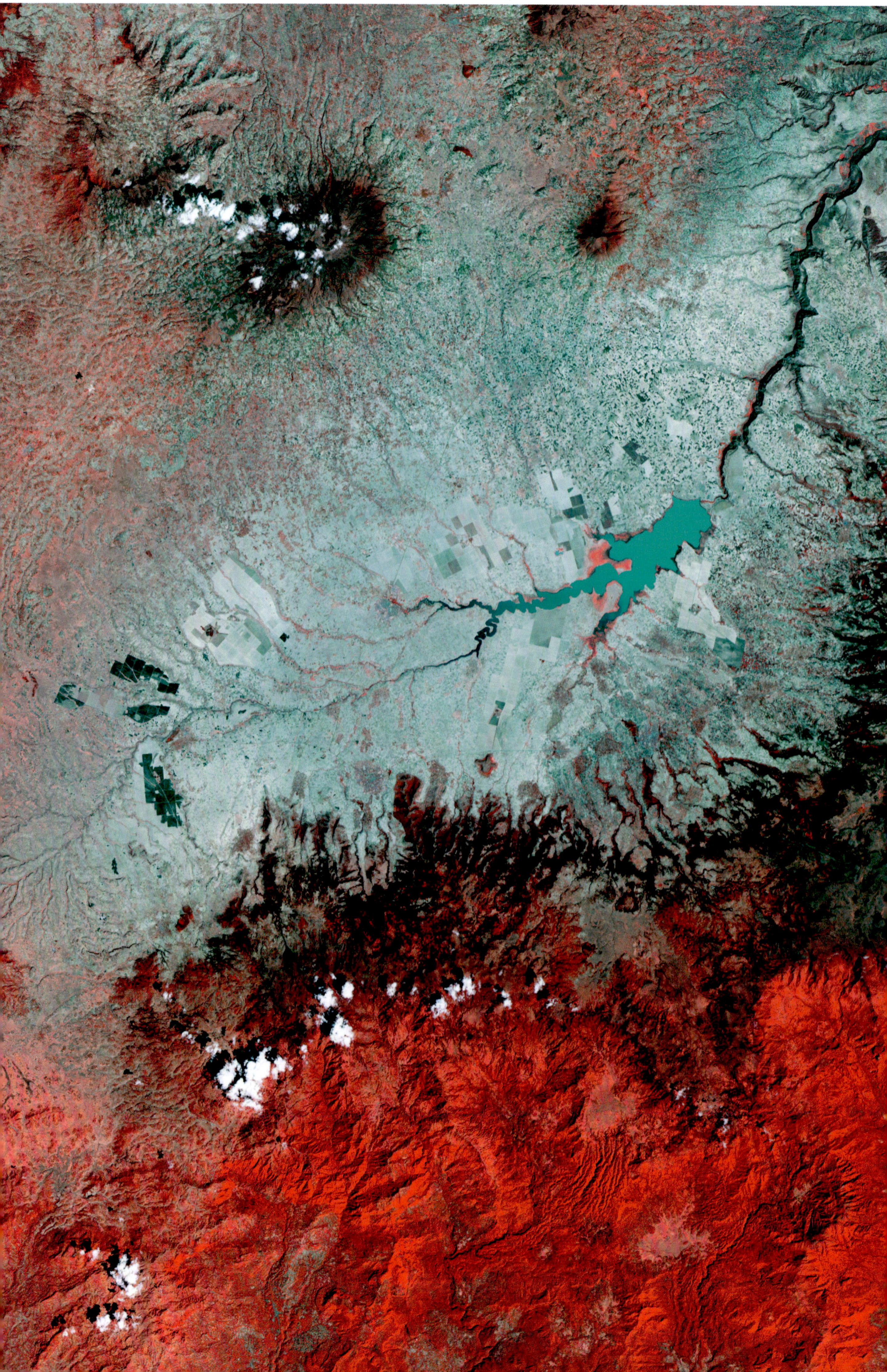

2. Cartographic methods

The coffee map layers were produced by combining high resolution satellite data (for forest cover) and niche modelling (for coffee suitability). Information on the use of the Atlas is found in Chapter 12 (Scope and use of Atlas).

To publish only the coffee areas and forest without any landscape features would render the Atlas near useless, as readers would find it difficult to determine their location. To make the map navigable and to give the reader context, we have over and under laid our coffee mapping data with infrastructure (e.g. roads and towns), boundaries, elevation data, rivers, lakes and protected areas (e.g. National Parks), and added graticules (the horizontal and vertical lines) for longititude and latitude on each Atlas page. The mapping layers used in the Atlas, and their origins, are given below:

Data set	Data origin
Main coffee areas	Royal Botanic Gardens, Kew – SCIP project
Coffee forest	Royal Botanic Gardens, Kew – SCIP project
Other forest	Royal Botanic Gardens, Kew – SCIP project
Other forest (north)	Global Forest Watch
Coffee niche	Royal Botanic Gardens, Kew – SCIP project
Coffee gardens	Royal Botanic Gardens, Kew – SCIP project
Coffee towns	Royal Botanic Gardens, Kew – SCIP project
Coffee localities	Royal Botanic Gardens, Kew – SCIP project
Protected areas	Environment & Coffee Forest Forum (ECFF)
Elevation data	SRTM 90 m Digital Elevation Database v. 4.1 (from CGIARCSI)
Towns	EthioGIS & Royal Botanic Gardens, Kew – SCIP project
Mountains	GEOnet Names Server
Airports	EthioGIS & Royal Botanic Gardens, Kew – SCIP project
Transport	EthioGIS, OpenStreetMap, & Royal Botanic Gardens, Kew – SCIP project
Borders	EthioGIS
Drainage and water bodies	EthioGIS & Royal Botanic Gardens, Kew – SCIP project

It should be noted that some of the data are based on outdated information, but this was the best we could access during the course of the project. Additional mapping data for towns, airports and transport (i.e. roads) was collected on-the-ground (during the SCIP project), and from satellite images (including review in Google Earth). The mapping data presented here should not be relied upon for detailed road navigation, and should be used with the following disclaimer in mind. *The information in this Atlas was derived from multi digital databases. A great deal of effort has been taken in the creation of this Atlas, but inevitably, omissions, or positional accuracy will occur. There are no warranties, expressed or implied, including the warranty of merchantability or fitness for a particular purpose, accompanying this product.* Notification of any errors would be appreciated.

The names of settlements (e.g. towns) and other mapped items are subject to change and regional variation. In addition, the names and geographical delimitations we have used for zones (see Chapter 12. Scope and use of Atlas) are unlikely to follow the most recent nomenclature and geographical boundaries precisely.

Colour scheme

Producing a colour scheme for the Atlas was a difficult task, as it needed to be readable and printable, to represent the features being mapped adequately, and to be aesthetically pleasing. We also wanted to account for readers who are colour blind. The colours chosen for the Atlas relate to the environmental suitability (the niche) for coffee, suitable forest cover, and other humid forest cover (presently not suitable for coffee). Please see Chapter 12 (Scope and use of Atlas), for further details.

These colours should allow the majority of colour blind readers (at least those with red-green colour blindness) to interpret the maps. To assist in this process we used colorbrewer2 (**http://colorbrewer2.org/**) to refine our colour representation for computer screens and for printing. A coffee bean symbol is used to represent the general distribution of coffee garden areas but not their precise locations, although the distribution is based on detailed computer modelling. Finally, we added a very lightly hued hill-shaded elevation layer to the background, to give the impression of topology. A topological map is provided in Figure 8; see page 61 for details.

Projection

To represent the land surface and distances as accurately as possible we used the same projection and datum for all of our mapping. Most of the data used were either in geographic coordinates (latitude and longitude) using World Geodetic System 1984 (WGS84) or Universal Transverse Mercator (UTM) projection, Zone 37N.

The national projection used for Ethiopia is UTM Zone 37 N using the Adindan datum. On testing the Adindan and WGS84 datum for on-the-ground difference, the maximum difference was 150 m, which is well below our mapping scale of 1:500,000. The Adindan and WGS84 datum are more or less interchangeable at the scale of this Atlas. The Atlas maps incorporate graticules for latitude and longitude in degrees and minutes. A summary of the projection used is given below:

Projection: Transverse Mercator	Scale factor: 0.9996
False easting: 500000	Latitude of origin: 0.0
False northing: 0.0	Linear unit: Meter 1.0
Central meridian: 39.0	Datum: WGS 1984

LEFT
Landsat 8 false colour composite satellite image, for West Arsi and a part of the Sidamo coffee area, south-east of Hawassa. Bright red represents forest and forest-like vegetation; lighter colours mainly represent cultivated and pastoral activities. See Atlas map page 12.
Image: Landsat 8. United States Geological Survey (2015).

Format

We have divided the Atlas into 40 A4 pages and have extended the mapping beyond the coffee areas into regions that might be of more general interest to the reader. An index map is presented on Atlas map page 1. To help with navigation, the adjacent pages of the Atlas are given in purple triangles at the margin of the page. The key is presented on the index map and, where space permits, on additional map pages.

Mapping of the humid (coffee) forest

A coffee forest map of humid forest cover (Transitional rain forest (TRF) and Moist evergreen Afromontane forest (MAF))[5] was produced from 38 Landsat 8 satellite images[6], covering 24 Landsat scenes (2013–14). The Landsat imagery was atmospherically corrected[7] and cloud masked[8] to generate a cloud-free composite image. From this image a Normalised Difference Vegetation Index (NDVI) layer was produced[9] and two thresholds derived to identify TRF and MAF. Forest cover (type and extent) was reviewed using expert assessment, including ground truthing during extensive fieldwork[10].

Mapping of coffee suitability

The coffee suitability (niche) mapping layers were produced using species distribution modelling (SDM), in conjunction with bioclimatic variables (BIOCLIM[11]). We trialled and examined all 19 BIOCLIM variables statistically, and used expert review to make sure the variables made ecological sense for Arabica coffee. Using a refined selection of BIOCLIM variables, we analysed the data using six modelling methods to produce a final ensemble model[12] for each side of the Rift Valley (east and west). We critically examined each model, keeping in mind the environmental and physiological requirements of Arabica coffee[10].

Humid (coffee) forest (Bonga, south-west Ethiopia) Image: Jenny Williams

3. Geographical overview

Location and size

Ethiopia is located in north-east Africa, in the Horn of Africa, north of the equator. The country extends from latitude 3°36'N to 14°53'N and longitude 33°00'E to 48°00'E, and is approximately 1,200 km from north to south, and 1,600 km from east to west, at the widest points. The land surface area is 1,104,300 km^2, making it the 10th largest country in Africa and the 27th largest country in the world by area. In comparison to other major coffee-producing countries, it is almost eight times smaller than Brazil (8,515,770 km^2), three times bigger than Vietnam (330,967 km^2), and about the same size as Colombia (1,141,749 km^2). Ethiopia is land-locked and bordered (clockwise, from the north) by Eritrea, Djibouti, Somalia, Kenya, South Sudan and Sudan.

Topology and elevation

Ethiopia is a country of great topological contrast. The western and central part of the country is dominated by the high plateau and mountains of the Ethiopian Highlands. The eastern part of the country is mostly dry lowland and semi-desert (Figure 5, page 28). The Ethiopian Highlands range from approximately 1,000–3,000 m, with numerous peaks above 4,000 m, including Ras Dejen (Ras Dashen) (c. 4,600 m) one of several high peaks in the Simien Mountains (in the north), and Mount Tulu Dimtu (4,389 m) on the Sanetti Plateau in the Bale Mountains (in the south). There are many other high peaks, including those above 4,000 m, which fall outside the Simien and Bale Mountains. The Ethiopian Highlands are divided by the Great Rift Valley, which runs roughly north-south; in the north-east the highlands are separated by the Afar lowlands. Elsewhere the Ethiopian Highlands are surrounded by lowland areas, except in the far north of the country where it borders with Eritrea (Figure 8, page 61). The highest point in Ethiopia is the peak of Ras Dejen (c. 4,600 m), and the lowest is in the Danakil Depression (125 m below sea level) of the Afar lowlands.

Major rivers and lakes

Ethiopia has numerous major rivers, which are important nationally and for bordering countries. Many of the larger rivers, such as the Blue Nile, are in deep valleys and form key features of the landscape. The main rivers are described briefly (below), anti-clockwise from the northernmost river first (Figure 5, page 28). The Tekezé River rises in the highlands of northern Ethiopia, east of Lake Tana, and flows north and then north-west to run along the border with Eritrea to Sudan, where it joins the Atbara River, a tributary of the Nile. The Blue Nile (known as the Abbay River in Ethiopia) starts at Lake Tana (see Atlas map page 34), and flows due south and then westward to Sudan. It meets the White Nile at Khartoum (Sudan), and together these rivers represent the two major tributaries of the Nile. The Baro River rises in the Ethiopian Highlands in south-western Ethiopia, and flows west through the Gambella Region (and through the city of Gambela; Atlas map page 16) to South Sudan, where it joins the Pibor River. To the south of the Baro is the Gilo River (Atlas map page 8), another major river of the Gambella Region, which originates in the southern part of the Ethiopian Highlands, from where it flows westwards to also join the Pibor on the border with South Sudan; the Pibor is a tributary of the White Nile. The Omo River is the principal river of southern Ethiopia, and rises in the central Ethiopian Highlands, runs due south, and empties into Lake Turkana on the border with Kenya. The Dawa River is situated in south-eastern Ethiopia, and after rising in the Sidamo highland area it flows south and then east to form part of the border with Kenya and Somalia. In Somalia it joins the Ganale Dorya River to become the Jubba River, which eventually empties into the Indian Ocean. To the north of the Dawa is the Ganale River, which rises and runs more or less parallel to the Dawa, until joining it at the border with Somalia. The Wabi Shabele River (Shebelle River) starts in the eastern Ethiopia Highlands, and flows in a south-easterly direction into Somalia towards the capital Mogadishu. Before reaching Mogadishu, it turns abruptly south-west, runs parallel to the coast, and then either dries up near the mouth of the Jubba River or (during substantial rainfall) reaches the Jubba River and empties into the Indian Ocean. The Awash River starts in the central highlands and is fed by the lakes of the Great Rift Valley. It flows in a north-easterly direction, and eventually supplies a chain of interconnected lakes that ends with Lake Abbhe on the border with Djibouti (Atlas map page 37).

Ethiopia has several major lakes and many smaller ones. Ethiopia's largest lake is Lake Tana, which is situated in the Amhara Region and has the region's capital (Bahir Dar) positioned on its southern shore. Lake Tana is fed by several permanent and many seasonal rivers. Most of Ethiopia's large lakes are concentrated in the Great Rift Valley; the major ones, from north to south, are Koka, Ziway, Langano, Abiyala, Shala (Shalla), Awasa (Hawassa), Abaya and Chamo. There are also some sizeable lakes in north-western Ethiopia, east of the city of Semera (Afar Region), including Lake Abbhe, Lake Gereri, Lake Afambo and Lake Adobed.

Geology and soils

The geology of Ethiopia is complex and varied. The oldest rocks (over 600 million years old) are Precambrian and comprise a wide range of sedimentary, volcanic and intrusive rocks, and these have been metamorphosed to varying degrees over time. Faulting and uplifting caused by tectonic movement is present across the landscape, and this has been accompanied by widespread volcanic activity, giving rise to huge volcanic deposits (magma/lavas and ash) and (igneous) intrusions. The youngest rock types are sedimentary (mostly Mesozoic and Cenozoic), including conglomerates, sands, clays and reef limestone. The sedimentary rocks are mainly a feature of eastern Ethiopia. Volcanic activity continues to the present day, with recent lava flows visible in the northern part of the Great Rift Valley (e.g. between Addis Ababa and Awash). Erta Ale, a

large shield volcano, is situated in north-eastern Ethiopia, in the northern part of the Danakil Depression, and is noted for its large, active lava lake. Most of the Ethiopian Highlands, the location of the major coffee-growing areas, are dominated by volcanic rocks (lavas and ashes) from the late Cenozoic. These rocks generally produce mineral rich, dark soils, of high agricultural value, which are well-suited to numerous crops, including coffee. Common soil types of the highlands include black soils and red lateritic soils. There are, however, a large number of other soil types in Ethiopia, with coffee farming occurring on many of these.

Natural vegetation, plants and other wildlife

Like many aspects of Ethiopia, the vegetation of Ethiopia is one of contrasts, from the hot, dry and sometimes desert-like lowlands, to the lush, cool forests of the Ethiopian Highlands. There are just over 6,000 plant species found in Ethiopia, of which around 10% are endemic, and 5,200 are flowering plants (angiosperms)[13]. In the *Atlas of the Potential Vegetation of Ethiopia*[5] the natural vegetation types of Ethiopia are mapped and described in detail. The following overview of Ethiopia's plant life is based on this work, and with the permission of the authors we have reproduced their vegetation map and classification for Ethiopia (Figure 5, page 28). Most of eastern Ethiopia is predominately low-lying, and, at least where natural vegetation still exists, is composed of desert and semi-desert scrubland (Desert and semi-desert scrubland (DSS)) and dry woodland and scrubland (*Acacia–Commiphora* woodland and bushland proper (ACB)). The western half of the country is dominated by the Ethiopian Highlands, the location of the country's main crop-lands. Where natural vegetation exists, the main vegetation types of the highlands are: dry evergreen forest, of shrubs and small and large trees, with grassland interspersed (classified as Dry evergreen Afromontane forest and grassland complex (DAF)); and cool, moist tropical forest (Moist evergreen Afromontane forest, and Transition rain forest (MAF and TRF)). These vegetation types typically occur between 1,000 m and 2,500 m. In the lower lying areas (400–1,800 m) between the higher elevation forest types, and in much of the western part of Ethiopia, a dense woodland-wooded grassland exists, characterised by small to medium-sized deciduous trees, interspersed with grassland (*Combretum–Terminalia* woodland and wooded grassland (CTW)). The Great Rift Valley area has its own vegetation type, characterised by mixed acacia woodland and grassland (*Acacia* wooded grassland of the Rift Valley (ACB/RV)); and in the lowlands of the Gambella Region, there is a specific type of wooded grassland (Wooded grassland of the Western Gambella region (WGG)). The mountainous areas of Ethiopia are home to two major vegetation types: the Ericaceous belt (EB) and the Afroalpine belt (AA). The Ericaceous belt (usually 3,000–3,200 m), is composed of shrubby or wooded heath-lands, and is dominated by the heath species *Erica arborea*. The Afroalpine belt (above 3,200 m) is mostly

High elevation wooded heathland of the Ericaceous belt, Simien Mts (northern Ethiopia) Image: Jenny Wlliams

LEFT
Gelada baboon, Simien Mts (northern Ethiopia)
Image: Jenny Wlliams

characterised by a sparse, open vegetation type with giant rosette plants, tussock grasses (and sedges), stemless rosette plants, cushion plants, and dwarf shrubs with small, tough (hard) leaves. There are other minor vegetation types, including those associated with freshwater lakes, marshes and swamps, salt lakes and salt pans (Figure 5, page 28).

Ethiopia has a large diversity of native (indigenous) animals. There are also many endemic species, i.e. species found only in Ethiopia, the most notable being the Walia ibex, the Ethiopian wolf, the grass-eating gelada baboon, and the mountain nyala. There are around 270 species of mammal, the most commonly seen include baboon, colobus monkey, grivet monkey (African green monkey) and warthogs. Other large mammals include hyena, hippopotamus, lion, African elephant, kudu, dik-dik, and Grévy's zebra. There are also numerous birds, with 863 species (including 19 endemics) recorded by Bird Life International, and more than 22,000 species of butterflies and moths.

People and culture

Ethiopia is a country rich in culture and history. Important historical and religious sites include the ancient city Aksum (from c. 400 BCE) with its impressive obelisks (stelae), tombs, castles, and churches; the monolithic rock-cut Christian churches of Lalibela (twelfth to thirteenth century); the cliff-face, rock-hewn churches of Tigray; and the imposing sixteenth and seventeenth century fortifications of Gondar. Aksum (Axum) is the locality of the Church of Our Lady Mary of Zion, which is believed to house the Ark of the Covenant, containing the Tablets of Law upon which are inscribed the Ten Commandments. Aksum also has a long-established history with Islam, as does the town of Negash, which is said to be the earliest Muslim settlement in Africa.

Today, Ethiopia has a population of around 100 million people, which is predominately rural and concentrated in the agriculturally superior highlands. In recent decades, urbanisation has increased steadily and is approaching 20 per cent of the population, in-step with cultural change and industrialisation. The country is ethnically diverse, with almost 90 different languages, the most widely spoken being Oromiffa (spoken by the Oromo), Amharic (by the Amhara), Somali (by the Somalis) and Tigrinya (by the Tigrayans). Amharic is the official language of central government and law, although Oromiffa is likely to become the second official language. English is the most widely used foreign language, although it is not widely spoken across Ethiopia. Arabic is spoken in some areas. Christianity and Islam are the two major religions, with a small percentage of traditional religions and faiths. There is a small but long-established Jewish community. Ethiopia is the spiritual homeland of Rastafarianism, or the Rastafari movement, which is centred on the town of Shashemene, although the history of the religion in Ethiopia is very recent (late 1940s). Agriculture, including pastoralism (keeping livestock), is the main source of employment and income generation for Ethiopia's population, employing around 80% of the population[14]. Services, manufacturing, and construction are among the main components of the industrial sector.

Food market in the ancient city of Aksum (northern Ethiopia)

Image: Jenny Wliams

Cultivated slopes with remnant *Combretum–Terminalia* woodland species (near Key Afer, southern Ethiopia) Image: Jenny Wlliams

Agriculture

Agriculture is the main driver of Ethiopia's economy, accounting for almost half of gross domestic product (GDP) and a large percentage of export earnings. Most of Ethiopia's agriculture, whether for subsistence or export, is undertaken by small-holder farmers. Principal crops include coffee, the narcotic khat, enset, pulses (peas, beans, lentils), oilseeds (e.g. sesame, niger, lupin, rape), cereals (teff, maize (corn), sorghum, wheat, barely, millet), potatoes, sugarcane, vegetables (including yams, sweet potatoes) and cotton. Coffee, followed by sesame and pulses, are the largest cash/export crops. The most commonly grown cereal (grain) is teff (*Eragrostis tef*), which is used to make injera, the traditional and widely consumed, fermented flat-bread of Ethiopia. Teff, wheat and barley are cool-climate crops, and generally grown above elevations of 1,500–2,000 m. Maize (corn) will withstand warmer temperatures but requires plentiful rain, and so is mostly limited to elevations similar to those that are suitable for the grains above. Sorghum and millet are more drought tolerant and can be grown in lower-lying areas. The narcotic crop khat (*Catha edulis*) is grown in many of the coffee-growing areas but is more drought and heat tolerant than coffee. The pulses and oilseeds are grown in various areas, but especially in the north, and are cultivated in a variety of areas depending on the specific crop species.

Enset (*Ensete ventricosum*), also called the false banana, is an important crop in the southern highlands, feeding an estimated 20 million Ethiopians. Despite looking like a banana its fruits are largely inedible, but the lower part of the stem contains large quantities of starch, which are made into a number of food products. Enset is also used as a source of fibres (e.g. for basket-ware), fodder, and as wrapping material for khat. Common vegetables include onions, peppers, potatoes, squash, and cabbage. Fruits include various citrus and strawberries, particularly from cooler areas, and bananas, avocados, and mangoes from the more tropical (hotter and wetter) parts of the country. In the more humid forest areas, honey, spices, timber and fuelwood are key products for household use and income.

Pastoralism and livestock production is a key part of the agricultural sector, with livestock including cows (of the Zebu type), which are the most dominant livestock animal, followed by sheep and goats. Equine animals (horses, mules, and donkeys) are also an important part of animal husbandry, being mainly used for transportation purposes. In the Rift Valley and adjacent areas and in eastern Ethiopia, camels also play a key role in transportation, while also providing milk and meat. Keeping poultry is very common in rural households, for consumption by the owners and for the local market.

Regions, zones, woredas and kebeles

Addis Ababa, located in a central position in the western Ethiopian highlands, is the capital city of Ethiopia. At an elevation of 2,200–2,500 m, Addis Ababa is the highest capital city in Africa. Ethiopia is divided into a hierarchy of administrative divisions, including regions, chartered cities, zones, woredas and kebeles. There are nine regions (capital towns are given in parentheses): Afar (Semera), Amhara (Bahir Dar), Benishangul Gumuz (Asosa), Gambella (Gambela), Harari (Harar), Oromia (Addis Ababa/Finfinne), Somali (Jijiga/Jigjiga), Southern Nations, Nationalities, and Peoples' Region [SNNPR] (Hawassa) and Tigray (Mekele). Addis Ababa and Dire Dawa are the two chartered cities. The regions are subdivided into zones (second-level division) as shown in Figure 8, page 61; Atlas map page 1 and listed in the Gazetteer (Atlas map page 43). Woredas (third-level subdivision) are generally subdivisions of zones, and each has its own elected council. Kebeles (fourth-level subdivision) are the smallest administrative unit and represent an administrative delimitation at the local level.

4. Coffee use and consumption

Coffee as a food and other uses

Arabica coffee has been used in Ethiopia as a food and beverage for many hundreds, if not thousands, of years. Its earliest use was no doubt restricted to gathering and eating fruits from trees growing wild in the forests, with semi-domestication and development as a food source probably coming soon afterwards. Coffee fruits are perfectly edible, and can be eaten raw, straight from the tree. When consumed in this way, the fleshy pulp (known botanically as the mesocarp) is eaten and the seed (the coffee bean) is rejected, or 'spat out', as when eating a cherry or similar fruit. However, some people chew and eat the raw bean, and used in this way it is sometimes referred to as 'chewing coffee'. The practice of drying coffee fruits for later use as chewing coffee is quite popular in Uganda (but using robusta coffee). Coffee pulp is very sweet and has a delightful flavour when ripe, qualities which make it suitable for its use as a food and beverage. The pulp is used fresh or, more commonly and conveniently, it is the dried husks (dried pulp) that are used. In the Arabian Peninsula (e.g. Yemen) and in Ethiopia, a beverage made from the husks is known as *qishr* (or *quishir*, *geshir*, *keshir*, *kisher*) but it may also be called *hocha* (or *hoja*) in Ethiopia. *Qishr* is quite popular in Hararge (Harar), where the husks are boiled in water for several minutes before the drink is served hot or slightly cooled. Hot or cold milk is sometimes added. Historians have suggested that *qishir* may have been the forerunner of coffee, particularly as it also known as *qewa* (or *kahwah*), which is thought to be linked to the origin of the name coffee[15–17]. Coffee husks are used to make a similar drink in Central and South America, where it is called *cascara*. More recently, but outside Ethiopia, processed coffee husks are being made into an additive for chocolate bars, and syrups for use as a sweetener or as a base for soft drinks.

A tea-like drink made from the leaves of Arabica coffee is a popular drink in parts of Ethiopia, and especially in Hararge (Harar), the south-west (e.g. Kaffa and Bench Maji), and the south (Bale). It is also popular on the Boma Plateau in South Sudan. In Oromia it is usually known as *coti* (or *kuti*), although in Hararge the names *qutti* or *quxxii* are used. The leaves

Coffee leaf tea, coffee husks and dried coffee fruits (from top to bottom); photograph taken in the city of Harar Image: Jenny Wlliams

LEFT
Ethiopian coffee ceremony (Mizan Teferi, south-western Ethiopia)
Image: Jenny Wlliams

are usually picked fresh from the trees, although sometimes fallen leaves are gathered, and either sun-dried or dry-roasted in a shallow pan. When required for consumption the leaves are either broken into small fragments or made into a coarse powder and used in a similar way to tea. Sometimes milk is added. *Coti* is also used as a medicine, for example in the treatment of headaches, fever, and stomach complaints. *Chémo* (or *chamo*) is a variant of *coti*, consumed in parts of Bench Maji and Kaffa. The leaves are roasted and powdered, as in the preparation of *coti*, but other ingredients are added, including ground chilli pepper, ginger, garlic, salt, local basil, and rue. *Chémo* is considered to be invigorating and strengthening, with some users drinking it to keep hunger at bay. It is often taken before a day's hard work and before long journeys, for example.

Traditionally, roasted coffee beans (seeds) are mixed with oils and fats, and with other substances, to make a variety of food items. In the Oromia region, and less commonly in Sidama, coffee is roasted and coated with hot butter and salt, to which spices and sometimes onions and garlic are added. This dish is known as *bunna qélla* (or *buna-quala*), and served slightly warm it is offered as a snack or appetiser, particularly during ceremonies or blessings. Indeed, there is even a specific ceremony based around the preparation and offering of *bunna qélla*. In a similar manner to *bunna qélla*, roasted and ground beans are mixed with butter or oil, shaped into balls and cooled for later consumption. Another variant is to mix the ground roasted beans with butter and sugar to make a paste, to be eaten with unleavened bread[18]. Roasted coffee is also used medicinally in Ethiopia. Tourists may be offered darkly roasted (partially carbonised) ground coffee powder as cure for diarrhoea or a sore throat. Less common is the use of the flowers for inflammation of the eyes, coffee pulp for the healing of wounds, and roasted coffee leaves for stomach ailments.

Roasted coffee as a drink

Despite many legends, and particularly that of Kaldi (Kahlid) the goatherd, and some dubious biblical references, there is no record detailing the date when coffee beans were first roasted and made into a hot beverage. Despite the strong links between Islam and Arabica coffee, there appear to be no records of coffee in the Quran (Koran).

The story of Kaldi the goatherd is the most famous legend concerning the origin of roasted coffee. The story goes that Kaldi was out one day tending his goats when he noticed they had become excited and frolicsome. On closer inspection he noticed they were eating the fresh, red fruits from some nearby trees. Kaldi ate these fruits and found they had a stimulating effect. Later, on the advice of his wife, he took the fruits/beans to a local holy man (or priest) to ask his opinion. The holy man was enraged with Kaldi, believing the fruits to be an evil force, and threw them into the fire. Some minutes afterwards, when they noticed a delightful aroma coming from the burning beans, the beans were removed them from the fire and water added to recapture the essence of the aroma. They found this to their liking, and coffee was born. There are many variants and elaborations of the legend, including those from Yemen. The story is said to originate from the ancient region of Kaffa, which was much larger that the modern-day Kaffa zone.

Ethiopian macchiato Image: Aaron Davis

The seemingly obvious etymological association of Kaffa and coffee, and the birth of coffee culture and drinking, is widely publicised, although this link appears to be without foundation[15]. The fact that coffee drinking is such an intrinsic part of Ethiopian society and culture certainly alludes to ancient usage. There is some evidence that the use of the coffee pulp to make beverages (see above) predates the use of roasted coffee beans, although the use of roasted (or at least cooked) coffee beans for food could also be linked to the origin of coffee. There are many early associations linking Harar (Hararge, or Hararghe) and the Yemen with the origin of roasting coffee beans to make coffee, with most evidence pointing towards somewhere between the thirteenth and the sixteenth centuries, and with a strong association with Islamic culture. Indeed, the Islamic link is clear: by far the most commonly used name for coffee in Ethiopia is *buna* (from the Arabica word *buun*), and the traditional Ethiopian coffee ceremony has many Islamic references. Detailed accounts of legends, historical usage and the spread of coffee drinking from Ethiopia to Arabia and the Orient, and beyond, can be found in numerous coffee text books[15–17,19].

Modern-day use of coffee

In modern-day Ethiopia coffee drinking is part of every-day life, and for many an important part of community and family living. The Ethiopian coffee ceremony (*tukke* or *bunna*) is well known, and although traditionally performed three times a day, modified and abbreviated version are often seen in hotels and restaurants. In the ceremony, coffee is roasted, made and served using specific traditional equipment, including a low wooden table or box-like table (*rekebot*) that holds the traditional ceramic coffee pot, the *jebenna* (*jebena*, *jabena* or *jebana*), small ceramic serving cups (*sinni*), and other items. The roasting part of the ceremony is undertaken using a charcoal burner (*medija*), and a shallow roasting pan (*yebuna mitad* or *metbeshia*), as used in most households and traditional coffee shops. The roasting coffee is turned frequently with a special wooden or metal spatula (*ye buna mequilia*), and after roasting is powdered using a wooden or metal pestle and mortar. The setting and ambiance are also important, including a covering of sedges (*qettema*) on the floor where the coffee is being made and roasted, and the burning of incense. The coffee is re-made and re-served two more times, each serving having its specific order, nomenclature and significance. The ceremony is performed by women and teenage girls. In most towns and villages, women will invite their neighbours for coffee to discuss the news, gossip, and catch-up on the activities of family and friends. A detailed account of the coffee ceremony is given by Tadesse (2017)[15]. Other types of coffee ceremonies are undertaken during births, marriages and other special occasions, with many regional variations.

Outside the home, in every town and village there are usually numerous coffee houses, providing either traditional or 'modern' espresso-style coffee. The traditional type is made in the characteristic fired-clay *jebenna*, and served in small handle-less cups, as in the coffee ceremony. The 'modern' type is made with an Italian espresso machine, to produce a large espresso-like (strong black) coffee, and served in a short glass or small Italian-style espresso cup. Many of the espresso machines have some age, and it can be a delight to see these beautiful old machines being used on a daily basis. A good helping of highly frothed milk is added to the espresso-like base, to make the Ethiopian version of the Italian macchiato, which is also called a macchiato. Sometimes the froth can be several centimetres high. Ethiopian coffee is almost universally dark to very darkly roasted, and ground to a very fine powder. Coffee-tea (*spreece*, or *chai-buna*) is very popular and is made by layering coffee on top of sweetened (and sometimes spiced) tea into a single glass or cup[18].

In Ethiopia, certain spices and herbs may be added to coffee, including rue (*Ruta graveolens*), cardamom (*Elettaria cardamomum*) and local cardamom (*Aframomum corrorima*), and less commonly ginger (*Zingiber officinale*), cloves (*Syzygium aromaticum*) and cinnamon (*Cinnamomum* species). In certain, places, such as Mizan Teferi, salt is either added or provided in a small spoon. Cow or goat milk may also be added to coffee; or a smaller amount of strong coffee may be added to a much larger volume of rich, warmed milk. Given the longstanding and popular coffee-consuming tradition it is hardly surprising that Ethiopia consumes more coffee per capita than any of the world's other coffee-producing countries. It is estimated that half, or perhaps as much as 60%, of the coffee grown in Ethiopia is consumed within-country.

Roasting coffee over charcoal

Image: Jenny Wlliams

A coloured engraving of Arabica coffee (*Coffea arabica*) from the book *An Historical Account of Coffee* by John Ellis (1710–1776), published in London, 1774.

5. Botany of Arabica coffee

Origin and genetic makeup

We will never know the precise details for the biological origin and evolution of wild Arabica coffee (*Coffea arabica*), but based on research on the plant and its close relatives it is possible to make some plausible inferences. From a wide range of scientific research we know that Arabica coffee is formed from the hybrid union of two African coffee species: robusta coffee (*Coffea canephora*) and Nandi coffee (*C. eugenioides*), or perhaps from two other very closely related species (extant or extinct). Robusta coffee has a wide distribution across much of tropical Africa; Nandi coffee is found mostly in the highlands of the central and northern part of the African Rift Valley area, and in some localities on the eastern side of the Rift Valley. It is not known where the hybridisation of these two species took place, but it is most likely that it occurred in a locality close to Ethiopia, or perhaps even in Ethiopia itself. Today wild robusta and Nandi coffee are found growing together, or close to each other, in South Sudan, the Democratic Republic of Congo, Kenya, Tanzania and Uganda[20]. However, because the climate has changed over recent geological time, the distribution of these two species is likely to be different today than it was many millennia ago. Indeed, robusta and Nandi coffee could have been present in Ethiopia at some time in the geological past.

The hybridisation of robusta and Nandi coffee was followed by a duplication of the number of chromosomes, a phenomenon known as allopolyploidy. In Arabica coffee the chromosomes doubled, so instead of having 22 chromosomes, as in all other coffee species, there are 44. Arabica is classified as an allotetraploid, because it has four sets of chromosomes. There are consequences associated with allopolyploidy; some good and some bad. Allopolyploidy can make a plant more vigorous and better able to adapt to a wider range of environmental conditions, as does hybridisation alone but without an increase in the number of chromosomes. Another consequence of allopolyploidy, although not universal, is the ability to self-fertilise. In self-fertilising plants, cross pollination (i.e. crossing with another plant) is not required for the fertilisation of the flowers, and thus for the production of viable fruits and seeds. This is a useful attribute for a crop plant, but it has a serious long-term consequence for both wild and cultivated plants. When plants self-fertilise they do not receive genetic material from other (genetically different) individuals, which over time can lead to a severe reduction in genetic diversity (inbreeding). This has probably been the case for wild Arabica coffee in Ethiopia, especially during the early period of its evolutionary life following the initial hybridisation event, when there would have been very limited opportunities for out-crossing. Moreover, available research infers that the hybridisation of robusta and Nandi coffee, and the formation of the allotetraploid Arabica coffee, may have happened only once[21], further limiting genetic exchange.

Despite these limiting factors, wild and cultivated Arabica coffee in Ethiopia possesses considerable diversity, and comprises the world's main storehouse of genetic resources for this crop species. This diversity has specific benefits for the coffee sector, locally, regionally and globally. Most directly, it provides an important source of new planting material for Ethiopian coffee farming, via seed and seedlings, and the intrinsic (genetic) variation that gives rise to the various flavour profiles found across Ethiopia. In the longer-term, protection of these genetic resources should be seen as a key element in the sustainability of the coffee sector in Ethiopia and world-wide.

Distribution of wild Arabica coffee

Arabica coffee was known to the scientific and medical community outside Ethiopia from at least the mid to late sixteenth century onwards. However, it was not until 1753 that it received its formal scientific name of *Coffea arabica*. It was named by the Swedish botanist, physician and zoologist Carl Linnaeus (1707–1778), which is why one sometimes sees the scientific name suffixed with 'L.', the abbreviation for Linnaeus (i.e. *Coffea arabica* L.). The plant was, of course, also well known outside the scientific community before the sixteenth century, especially in Ethiopia and Yemen, and was already being grown and consumed outside Africa and the Arabian Peninsula. Carl Linnaeus, and botanists who came before him, and even those working up until the early and middle part of nineteenth century, believed that the coffee plant came from Yemen (Arabia felix; or in Latin *Arabia felici*), hence the name 'Arabica'. It was not until the mid to late nineteenth that it dawned on the scientific community that this was actually a plant from Africa, and only much more recently that its true biological homelands are the forests of upland Ethiopia and South Sudan[1]. Robusta coffee (*C. canephora*) only became known to the scientific world relatively recently, from around the turn of the twentieth century, although its cultivation and use (as a chewing coffee) in Uganda was recorded in the 1860s, and like Arabica coffee, it has probably been used by local people for millennia.

There are 124 species formally recognised as belonging to the coffee genus (*Coffea*), and they are found throughout Africa, Madagascar and other western Indian Ocean Islands, Asia and Australasia[20,22]. In Ethiopia two species occur naturally (i.e. grow wild): Arabica coffee (*C. arabica*) and *C. neoleroyi*. *Coffea neoleroyi* is a small deciduous shrub or treelet, occurring in seasonally dry forests (in *Combretum-Terminalia* woodland and wooded grassland (CTW))[5] in south-western Ethiopia, at elevations of 600–750 m. It also occurs in eastern South Sudan, and northern Uganda. *Coffea neoleroyi* is a rare species, seldom seen and little-known, even by botanists, and is not used in the production of coffee. Arabica coffee (*C. arabica*) grows wild in Ethiopia and South Sudan, although in South Sudan it is restricted to the Boma Plateau, close to the border with Ethiopia. Mount Marsabit in northern Kenya is often given as a location for wild Arabica, although research indicates that the Arabica plants growing on this mountain probably originated from a cultivated source, and that the environment and location is not consistent with natural occurrence[1]. Further research at Mount Marsabit is warranted, in order to fully assess the status of Arabica plants in this locality. Various literature sources give the Imatong Mountains in southern South Sudan as a location for wild populations of Arabica coffee, but the only known record from these mountains is most likely to be robusta coffee (*C. canephora*), which occurs wild in the Imatongs.

The largest and most diverse populations of wild Arabica occur in the forested upland areas of south-western Ethiopia[23] (Figure 1). The populations are found in the coffee areas of Jimma-Limu, Kaffa, Bench Maji, Tepi, Illubabor, and the western part of Wellega (Figure 1; and Figure 7, page 49). There are also wild populations on the eastern side of the Rift Valley, but these are restricted to forested areas of the Bale coffee area (Figure 7, page 49) and most notably Harrena forest. It has been suggested that the populations in Bale are not indigenous (i.e. they were not originally wild in this area), but were brought there from other localities on the eastern side of the Rift Valley, possibly from Sidamo. Research using molecular (DNA) techniques, however, suggests that Arabica is part of the natural vegetation, and this is supported by the fact that the forests in the Bale Mountains have a plant species composition that is very similar to the humid (coffee) forests of south-west Ethiopia.

In Ethiopia, Wild Arabica is found in nearly all the forested areas with excellent or good coffee (niche) suitability, which are coloured dark green on the Atlas maps. Exceptions include the western part of the Sidamo coffee area, which are mostly covered by dense agroforestry. However, it could be that some or most of the coffee currently being grown in the Sidamo coffee-growing area originally came from wild forests that were once present in the area. Maps showing recorded locations for wild coffee in Ethiopia are found in Davis *et al.* (2012)[1].

Ecology

In Ethiopia and South Sudan, wild Arabica coffee occurs as a small understory tree in moist (humid), evergreen montane forest, at elevations from 950–1,950 m, although 1,200 m is a now a more realistic lower elevation limit, mainly due to the loss of forest at lower elevations. Wild Arabica coffee occurs in Moist evergreen Afromontane forest (MAF), and Transitional rain forest (TRF)[5] (Figure 5, page 28). MAF and TRF forests are composed of a diverse array of evergreen trees and shrubs, and represent what is best described as cool-tropical rainforest. MAF and TRF forest types, sometimes referred to as 'coffee forest', are mostly restricted to the south-eastern and southern part of the Ethiopian Highlands, as shown in Figure 1. Their potential area of cover would have been much greater than it is at present, as much of this forest type has been removed to provide land for agriculture and settlement (see Chapter 7). The potential forest area for MAF and TRF, i.e. where these forest types could grow if it were not for agriculture etc., is shown in Figure 5 (page 28); the actual remaining area for these forest types (collectively termed humid forest) is shown in Figure 1, and in the Atlas pages (Map pages 1–41). Arabica coffee is predominately cultivated in MAF and TRF areas, but in some locations, particularly in the Harar coffee zone (Arsi, West Hararge, and East Hararge), the Amhara Region, and in highland areas in south-eastern Ethiopia above c. 2,000 m, it is cultivated in the wetter areas of the high elevation evergreen forest (Dry evergreen Afromontane forest and grassland complex (DAF); Figure 5, page 28).

Wild Arabica coffee plays a key role in coffee production in Ethiopia, not only as a harvested crop in forest and semi-forest production systems, but also as the source of seeds and seedlings for the planting and re-planting of many of Ethiopia's coffee-production areas. The coffee forests of Ethiopia also have immense value as the main storehouse of wild Arabica coffee genetic resources, which have been used as a source of disease-resistant cultivars and for improving the diversity of the species cup profile (e.g. unusual and superior taste profiles). The genetic diversity of Ethiopia's wild Arabica populations far exceeds that of cultivated variants of Arabica coffee used in crop production elsewhere in the world.

Indigenous populations are often highlighted as a key resource for the medium- to long-term sustainability of global Arabica production.

The humid evergreen forests of Ethiopia are recognised as a globally important hotspot for biodiversity conservation, and are part of the Eastern Afromontane Biodiversity Hotspot[24], one of 34 globally important priority areas for biodiversity conservation. Given that over 50% of the Afromontane region higher than 1,500 m and about 80% of the Afromontane region surpassing 2,000 m is found within Ethiopia, the 'coffee forests' and other traditional coffee production systems in the highlands of Ethiopia are important for biodiversity conservation, both nationally and globally.

Morphology and development

Tree shape and architecture

Arabica coffee is a small evergreen tree (not a shrub), capable of reaching 7 m in height, with a 3 m spread, or thereabouts. At maturity the Arabica tree is more or less triangular in outline, and conical in its three-dimensional shape, due to it having the longest (main) branches at the base of the tree, which progressively shorten towards the top. There is a single upright (orthotropic) main stem (the trunk), the diameter of which varies according to age and growing conditions, from 2–3 cm on young trees to 4–8 cm on older plants, when measured at the base of the tree. On very old wild trees the stem can be 8–13 cm in diameter[25], and in rare cases 15–20 cm[26]. The main branches arising from the trunk may also be termed 'laterals', 'primaries', or 'primary branches'. The main branches are more or less regularly spaced up the main stem, arranged in opposite pairs, and held either at right angles to the main stem (horizontal or plagiotropic) or at low angles (near-horizontal, up to 30°, or sometimes more like 45°). The

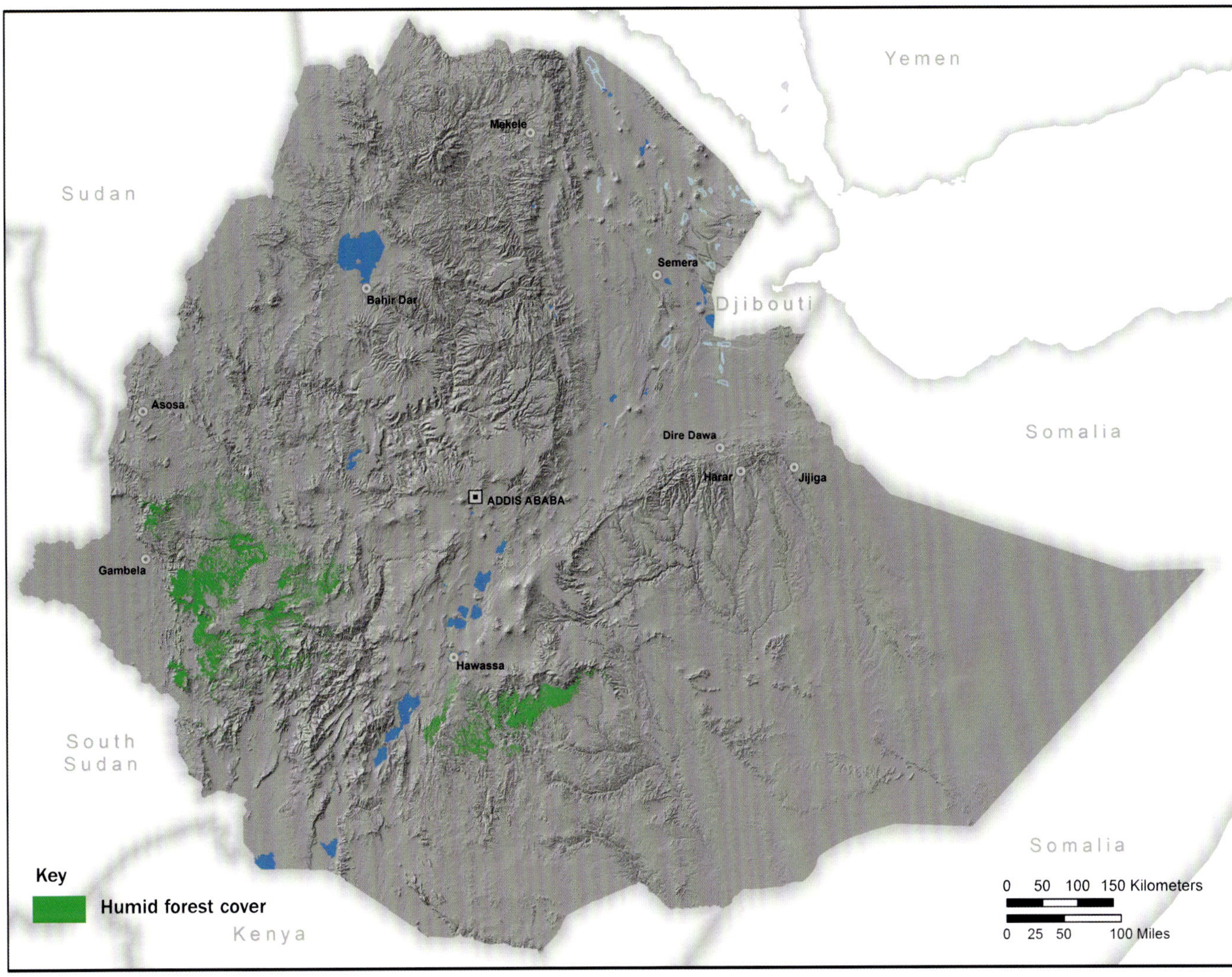

Figure 1. Humid forest cover (2014–15) derived from Landsat 8 imagery. See Chapter 2 for details. Some of the humid forest cover is represented by dense agroforestry (particularly in the western part of the Sidamo area, south of Hawassa).

primary branches give rise to oppositely arranged secondary, tertiary and sometimes quaternary branches, which are often collectively referred to as 'sub-laterals'. If the tree is pruned, 'topped' (cut off at the top), or 'stumped' (cut down at the base), or if the plant is not growing properly, the shape will be altered. Moreover, with age of 10–20 years or more, the Arabica plant develops into an untidy, leaning tree, with a crooked and/or twisted trunk, losing its regular, more or less conical shape and regular branching pattern.

Considering that many trees live for hundreds of years, and even more than a thousand years, Arabica coffee trees are ranked as short-lived. In general, farmed Arabica trees start to cease maximum productivity after 10–15 years, unless stumped or heavily pruned, and live for a total of 20–30 years.

Annual tree ring counts and other evidence, such as the diameter of the main trunk (see above), suggests that they can live for up to 100 and perhaps even 140 years[15] in very exceptional circumstances. However, it seems likely that these longer-lived Arabica trees would have been stumped numerous times. Wild Arabica trees usually grow much slower and may live longer than farmed trees.

Roots

The following description of roots is largely based on the work of Nutmann[27–29]. The roots of the Arabica coffee plant comprise a tap root, axial roots, lateral roots, (main) permanents, feeder bearers and feeders. The tap root, which is usually branched, extends downwards from the trunk for about 30–45 cm. From the tap root run four to eight axial roots, which grow downwards into the soil for 2–4 m. Also attached to the tap root are the lateral roots, which either run more or less horizontally through the soil to form the main plate of roots, or run downwards and branch in many directions. The permanents are the smaller, woody (cutinised) roots (c. 3 mm in diameter or somewhat wider) arising from the axial and lateral roots, and comprise the main framework of the root system. The feeder bearers, which are also woody, are rather evenly spaced along the permanents, and bear the feeders. The feeders, which are responsible for water and nutrient uptake, are white and non-woody (at least for around three years) and mostly less than 1.5 mm in diameter. The bulk of the feeders are usually 30–40 cm from the soil surface, but this can vary depending on many factors, and particularly on the soil type and water availability:

in rich deep soils they are deeper; in poorer soils, mulched soils (natural or through farming practices) and irrigated soils, they are mostly closer to the soil surface[17].

Leaves

The leaves are oppositely arranged, i.e. the leaf stalks (petioles) and thus the leaves are held in opposite pairs on the shoots. Situated between the petioles are small bract-like structures called stipules, or more precisely interpetiolar stipules, and these are more or less triangular in shape. The leaves are evergreen but short-lived; each leaf only staying on the plant for about one year, or sometimes longer. All parts of the leaf lack hairs (i.e. they are glabrous). Leaf size is highly variable, depending on age, the environment and genetic origin, but is usually between 7–18 cm long and 3–9 cm wide. The outline shape of each leaf is more or less elliptic (broadest at the middle and tapering towards the ends), and at the tip there is a small but distinct blunt point (the acumen). The margins of the leaves are straight, and neither toothed nor indented in any way; leaves of this type are termed as 'simple'. Each leaf has a midrib (main vein), 6–10 pairs of secondary veins (nerves), and numerous finer tertiary veins. The leaves are rather thin at first but with age become somewhat leathery (subcoriaceous). The upper surface of the leaf is dark green, smooth and shiny, and very slightly waxy; the lower surface is light to medium green, and matt (not shiny). On the lower surface of the leaf, in the axils between the midrib and the secondary veins are small, swollen, pit-like structures called domatia, although these are difficult to see with the naked eye.

Prominence has been given to the colour of young leaves as one of the key recognition characteristics for two important early coffee variants (often considered as varieties of *Coffea arabica*), 'typica' and 'Bourbon'. The 'typica' variant has bronze young leaves and leaf tips, whereas in 'Bourbon' these are green. Examination of wild and farmed Arabica coffee in Ethiopia shows that this character is highly variable, and that both types can be found growing in close proximity.

Flowers

The flowers are positioned in the axils of the leaves, i.e. between the (lateral) shoot and the stalk of the leaf (the petiole). There are one to ten flowers in each leaf axil, that is, 2–20 in a cluster (i.e. per leaf pair), which are grouped together in small numbers in structures known as inflorescences. Arabica coffee plants growing in forest or deeply shaded places generally produce fewer flowers, as few as two per cluster. Each lateral shoot produces 5–10, or sometimes up to 12 clusters of flowers depending on the age of the plant, its health and its genetic characteristics. When growing well, each Arabica coffee tree will produce tens of thousands of flowers.

In rain-fed conditions, the dormancy of the flower buds is broken by the first rains of the wet season, but this must be preceded by a dry (resting) period of 3–4 months. Flowering usually occurs 8–12 days after the rains, depending on temperature and other factors. The flowers generally open in the morning and fade by midday or the afternoon, either falling to the ground soon afterwards or staying on the plant for some days in a dry, brown state.

The main part of the flower (the corolla) is white, and is united at the base to form a tube-like structure (the corolla tube), upon which are five (sometimes four or six) lobes (the corolla lobes). Inside the corolla there are five stamens. Each stamen has an anther sac (the structure that houses the pollen), which is held on a fine stalk (the filament); the other end of the filament is attached to the inside of the corolla tube, at the throat (the uppermost part or the corolla tube), between and just below the lobes. The stamens are often referred to as the 'male' parts. In the centre of the flower is the long, slender style, which is terminated by two style lobes, the (stigmatic) surfaces of which receive the pollen. At its base, the style goes through the centre of a structure known as the disc, and connects to the ovary. Each ovary has two compartments (the locules), which each contain an ovule. Once fertilised by the pollen, the ovules develop into the seeds (the coffee beans). The style and ovary are often referred to as the 'female' parts of the flower.

LEFT
Young tree of Arabica coffee (Gore, south-west Ethiopia)
Image: Jenny Williams

Flowers of Arabica coffee (Bonga, south-west Ethiopia) Image: Aaron Davis

Fruit and seed

The flowers are either pollinated by bees or undergo self-fertilisation, as discussed on page 15. After pollination the fruit starts to develop, initially from a small green, round structure of a few millimetres in diameter, through various stages of ripening to the mature fruit (known botanically as a drupe). The ripe fruit, which is commonly referred to by those working in coffee as the 'cherry', is usually red, rarely yellow, and smooth apart from the remains of the fruit stalk at the base and the disk (formerly the disc of the flower; see above) at the very top. The mature fruit is made up of several layers. On the outside is the fruit skin (the epicarp), inside this thin layer is the watery pulp (the mesocarp), and then the hard, crunchy layer known as the parchment or hull (the endocarp), which surrounds the green coffee bean (the seed). The seed is covered by the silver-skin (the seed testa), a thin layer of cells that turn almost paper-like when dry. Inside, the seed is composed mainly of storage cells (the endosperm), which provide nutrition for the tiny embryonic plant (embedded within the endosperm at the base of the seed) mainly in the form of carbohydrates, lipids and protein.

In total, the coffee fruit takes 8–9 months to ripen. In the first 6–8 weeks after flowering (and fertilisation) the fruits remain small and round and are referred to as 'pin-heads'. Following this, from 8–18 weeks there is a rapid growth phase, when the internal parts of the fruit undertake rapid development and the fruit increases steadily in size. There is then a slow growth phase, which lasts for around 2 weeks[30]. From about week 18 to week 32 fruit development continues, especially the development of the endosperm and the accumulation of dry matter. Finally, from week 26 to week 36 the pulpy part of the fruit (the mesocarp) swells and becomes watery, the fruit skin (the pericarp) turns from green to yellow, and then finally red at maturity. The speed of development depends on numerous environmental factors, including rainfall, temperature, soil, sunshine hours and light intensity. Generally, the fruit ripening stage takes around 10 weeks, during which the sugar content increases markedly.

The seedling and its development

Under natural conditions, animals, including monkeys, birds, and civets (for fallen fruits), disperse the seeds within forests and between forest patches. The sweet and nutritious fruits are ingested by the animals, a process which removes all the pulpy parts of the fruit leaving the seeds intact. The seeds are voided on to the ground within the faeces of the animal, ready for the rains and then germination. Soon after germination the seed is pushed above the soil on the young stem, at which stage it is referred to as a 'soldier'. Inside the above-ground seed are the first two young leaves (the cotyledons), which continue to draw reserves from the endosperm and eventually (after 3–4 weeks) force off what is left of the seed (mostly the remains of the parchment). The cotyledons are vaguely ear-shaped, and therefore look very different from the normal leaves. Shortly afterwards the true leaves appear and further leaf pairs develop. From this stage the young plant will grow and develop to maturity. In cultivation, maturity is defined by the first crop, which takes around 4–5 years from germination. In the wild, growth is usually much slower, mainly due to competition with other plants; the fruiting stage might not be reached until a decade and perhaps even longer, after germination.

Ripe and ripening fruits of Arabica coffee (Harenna Forest, Bale Mts, southern Ethiopia)
Image: Jenny Williams

6. The coffee climate

Arabica coffee grows successfully, and produces a marketable crop, within a narrow range of environmental conditions. The interplay between rainfall, temperature and seasonality are the main factors governing coffee plant growth, including flowering and fruiting stages. Coffee will become increasingly stressed as the air temperature increases and the soil moisture decreases (due to lack of rainfall), and *vice versa*, within the optimum climatic range for the species (see below). Temperature and rainfall are strongly influenced by seasonality, i.e. the annual weather patterns, as indicated in Figures 2, 3 and 4.

The seasons

The climate of Ethiopia's coffee-growing areas is tropical, but because they occur in highland localities (above 1,000 m) they can be classified as cool-tropical. The seasons are largely defined by rainfall. There are three main seasons: Bega, Belg, and Kiremt (or Kremt).

Bega (October to January/February) — the long dry season. This is when coffee harvesting and processing take place. The severity of the main dry season depends largely on location but also on other physical characteristics, including elevation, slope and aspect. There may be slight to moderate rainfall or barely any at all.

Belg (February to May) — the first rains before the main wet season, which may either represent the early part of the wet season or form a distinct short wet season followed by a short dry period before the main rains. This is the main period for coffee flowering, fruit initiation and early fruit development stages.

Kiremt (Kremt) (June to September) — the main wet season. In areas where the Belg period is distinct from this main wet season, the rainfall pattern is bimodal (east of the Rift Valley); where the bulk of the yearly rainfall is confined to Kiremt, the rainfall pattern is unimodal (west of the Rift Valley). This is the period for the middle and later stages of coffee fruit development. Around the end of September there is a sharp decline in rainfall, as Kiremt comes to an end and the dry season (Bega) begins. During the later part of this season, into the early part of the dry season, is when coffee fruit ripening occurs.

Throughout the coffee landscape, temperatures are generally higher during the dry season and lower in the wet season(s). Seasonal rainfall and temperature profiles for six coffee-growing locations are shown in Figure 2.

Rainfall (precipitation)

Rainfall is the most important factor governing the distribution of coffee farming and wild coffee forests in Ethiopia[10,31,32]. The distribution of rainfall varies greatly across Ethiopia (Figures 2 and 3), according to the season, location, elevation and physical features of the landscape. Clear annual patterns are evident, although rainfall is extremely variable.

The highest rainfall is on the western side of the Rift Valley, in the main South West coffee zone, with the western-most part of the zone generally being wetter. The coffee areas of the south-west typically receive 1,500–2,100 mm of rain per year. In the southern part of the South West coffee zone there is generally less annual rainfall (1,300–1,700 mm per year). The North coffee zone, including the Amhara and Benishangul Gumuz coffee areas, receives 1,400–1,800 mm per year, although the main rains come later in the year and end earlier, compared with the South West coffee zone. This results in a shorter, more intense wet season and a longer dry season (see Bahir Dar, Figure 2).

On the eastern side of the Rift Valley, the Harar coffee zone usually receives lower rainfall than all other coffee areas (800–1,200 mm per year). The Sidamo coffee area in the south is wetter (1,200–1,600 mm per year), and the Bale coffee area somewhere in-between (1,000–1,400 mm per year).

Air temperature

The ideal average minimum temperature for Arabica coffee is 12–14°C, (Figures 2 and 4); temperatures at or slightly above 0°C can be tolerated only for very short periods (e.g. a few minutes to a few hours). The ideal average temperature is 18–22°C, and ideal average maximum (daytime) temperature 25–27°C. Maximum temperatures of 30–32°C are not lethal if they exist for short periods (hours), and if there is sufficient water available in the soil. Extreme temperatures for coffee beyond 32°C up to 34°C, and perhaps even a high as 38°C, can be tolerated for very short periods (a few hours), but only if there is adequate water in the soil. If the soil is not wet enough, then the extreme maximum temperatures for coffee plant survival will be lower; for example, temperatures around 30°C could be lethal within less than 24 hours if the soil is very dry.

Temperature patterns in Ethiopia vary according to the season, location, elevation, and physical features of the landscape. In the coffee areas on the western side of the Rift Valley, the minimum temperature is typically 11–14°C, the average 15–20°C, and the maximum 23–27°C. In the coffee areas on the eastern side of the Rift Valley, the minimum is 9–13°C, the average 16–20°C, and maximum 24–26°C. Elevation has a direct and almost uniform relationship with air temperature, with an approximate decrease of 0.6°C for every 100 m increase in elevation. Nevertheless, other physical factors (e.g. slope and aspect), weather (e.g. rain, wind), and vegetation (e.g. forest cover) will also influence temperatures, depending on location and time of year.

Humidity and other factors

In the coffee areas of Ethiopia, humidity is mainly linked to rainfall and temperature. In the dry season (October to January/February) the monthly average relative humidity (RH%) of a typical coffee-growing area within the South West zone is around 55–80 RH%, and in the wet season (June to September) 90 RH% or higher. There is a considerable difference in humidity between day (lower RH%) and night (higher RH%). The amount of sun (sunshine hours) is linked to seasonality, with greater cloud cover and less sunshine in the wet season(s), and less cloud and more sun in the dry season.

Climate change

It is clear from historical weather station data that the climate of Ethiopia has changed. In particular, there is good evidence that temperatures have increased steadily and significantly throughout Ethiopia, as well as indications that rainfall has decreased in some areas, including some of the main coffee areas[10,32]. Feedback from coffee farmers and coffee-farming communities supports recorded climate data, as does observation of coffee farms and Arabica plants occurring in climate-sensitive areas. Projections for climate change across this century are not encouraging, with temperature increases projected to continue at a steady and significant rate. Changes in rainfall are less certain, with projections ranging from a slight increase to a slight decrease, or little change, throughout this century[10].

These changes will have a substantial negative influence for coffee production in Ethiopia, due to declining climatic suitability in many of the areas that currently grow Arabica coffee (Figure 7, page 49). In terms of a gross simplification, in places where it is too hot and too dry to grow coffee (including the low elevation areas classified as Unsuitable, Marginal and Fair; see Figure 6, page 29 and Atlas map pages) coffee farming suitability is projected to deteriorate (worsen); however, many areas that are currently too cold for coffee farming could come into better suitability for Arabica coffee (particularly at higher elevations close to areas that currently have Good and Excellent suitability); and many (but not all) areas that have Excellent suitability should be more climate resilient.

Despite the negative outlook, there could be the potential to substantially increase coffee production in Ethiopia, if new coffee farming areas are created, particularly at higher elevations where it is cooler, and often wetter. Fortunately, Ethiopia is blessed with an abundance of high ground (over 2,000 m), providing substantial potential for so-called 'up-slope' migration to cooler areas. There could be the potential for a fourfold increase, although a doubling of the suitable land area for growing coffee might be more realistic given all the various constraints[10].

Transitioning in this way would clearly involve some major changes: new coffee production areas would have to be established, during specific time periods and in specific locations[32]. There are indications that this transition is already taking place, with coffee now being grown at elevations previously not seen before (Chapter 10). In short, there could be a route to continued and even expanded coffee farming in Ethiopia, even under climate change, if careful and timely interventions are made. There could also be a positive environmental benefit, as many of the new coffee areas would require significant tree planting in order to provide the right conditions for long-term coffee cultivation.

A non-technical overview of coffee and climate change in Ethiopia is provided in the report *Coffee Farming and Climate Change in Ethiopia: Impacts, Forecasts, Resilience and Opportunities*[32]. The science on which the report is based can be found in our associated research paper[10]. The Atlas, research paper[10] and summary report[32] can be used together as a base-line resource for scoping local and national climate resilience strategies for the Ethiopian coffee-farming sector.

View from Mt Tulu Dimtu (4,389 m), Bale Mts (southern Ethiopia), with Afroalpine belt vegetation
Image: Jenny Williams

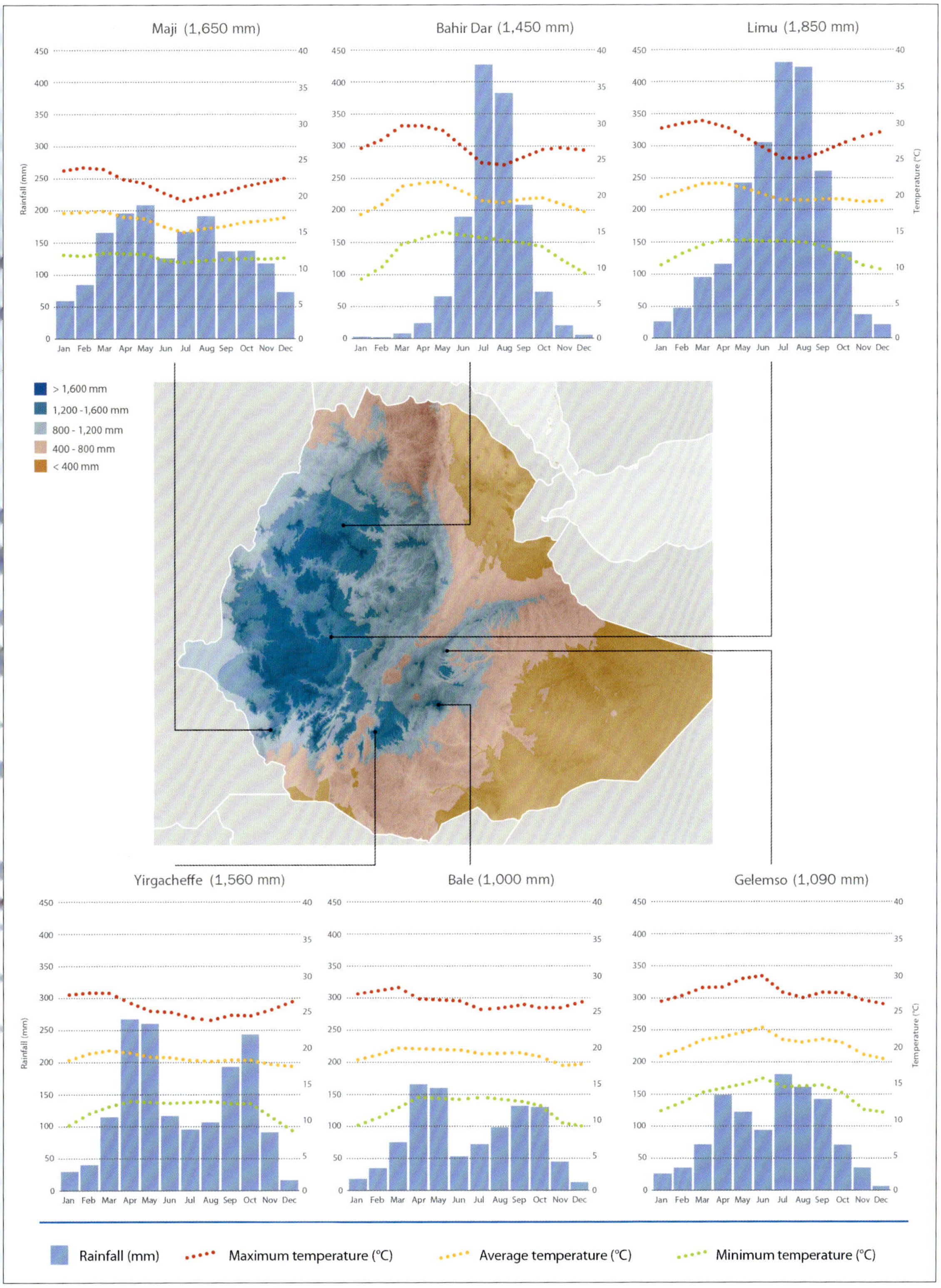

Figure 2. Annual rainfall map of Ethiopia (centre), and rainfall and temperature profiles for six coffee-growing locations. Approximate total annual rainfall for each locality is given in parentheses. Bimodal rainfall pattern (east of Rift Valley); uni-modal rainfall (west of Rift Valley). Graphs based on historical climate data (30-year averages) and WorldClim[11]. Yearly rainfall map of Ethiopia from WorldClim[11].

Bahir Dar. (Zege Peninsula). Sufficient annual rainfall, but the dry season (October to February/March) is very dry and coincides with the highest temperatures of the year. The bulk of the rain falls in the main wet season (May/June to September). Coffee suitability is Marginal (coffee-growing is possible but often problematic, with poor and very inconsistent yields) and is often negatively influenced by drought during the dry season. **Limu**. Sufficient rain throughout the dry season (October to February) and good rains in the main wet season (May/June to September). Highest temperatures coincide with sufficient rainfall. Coffee suitability is Excellent (coffee production good to excellent, with the potential for consistent, high-quality yields). **Maji**. Sufficient rain throughout the dry season (October to February) and good rains in the main wet season (May/June to September). Highest temperatures coincide with sufficient rainfall. Coffee suitability is Excellent (coffee production good to excellent, with the potential for consistent, high-quality yields). There is a slight bimodality to main season rainfall pattern. **Gelemso**. Barely sufficient rain throughout the dry season (October to February) and low rainfall in the main wet season(s) (April to September). Highest temperatures may not always coincide with sufficient rainfall. Coffee suitability is rated Unsuitable to Marginal (coffee-growing is possible but often problematic, with poor and inconsistent yields) especially without intervention (e.g. irrigation). **Bale**. Barely sufficient rain throughout the dry season (October to February), and low rainfall in the main wet season(s) (April to September); low rainfall is offset by slightly lower temperatures. Coffee suitability is Good to Excellent (coffee production good to excellent, with the potential for consistent, high-quality yields). Yield and quality may be negatively affected in some years (by adverse weather conditions, especially low rainfall). **Yirgacheffe**. Sufficient rain throughout the dry season (November to February) and sufficient rainfall in the main (and slightly extended) wet season(s) (April to October). Highest temperatures usually coincide with sufficient rainfall. Coffee suitability is Good to Excellent, with the potential for consistent, high-quality yields. Yield and quality may be negatively affected in some years (by adverse weather conditions, especially low rainfall). For further interpretation see Figures 3–8.

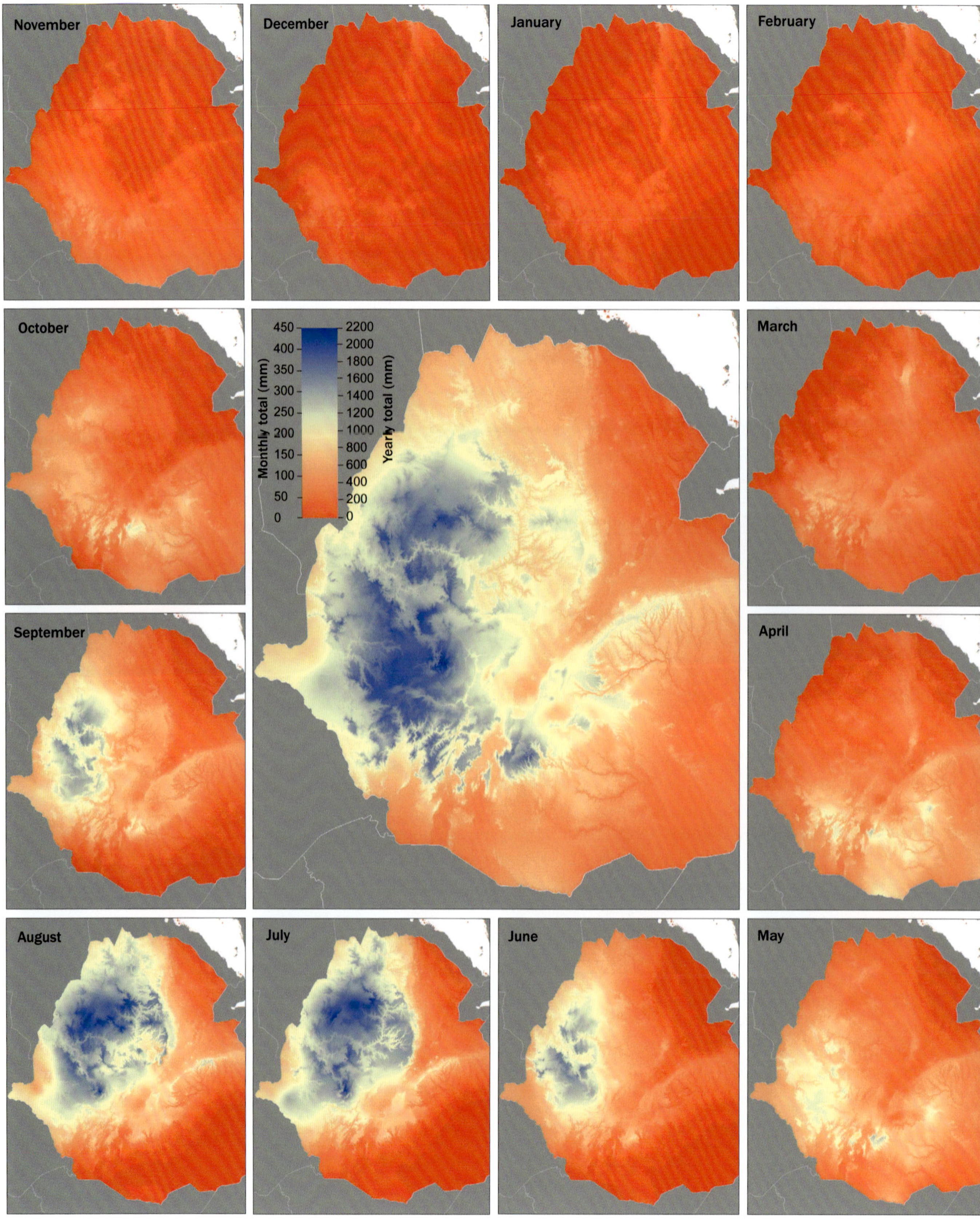

Figure 3. Total monthly rainfall (mm) distribution from January to December (clockwise); see central map (top left hand side) for key to rainfall amounts (mm/month). Total yearly rainfall (mm): central map.

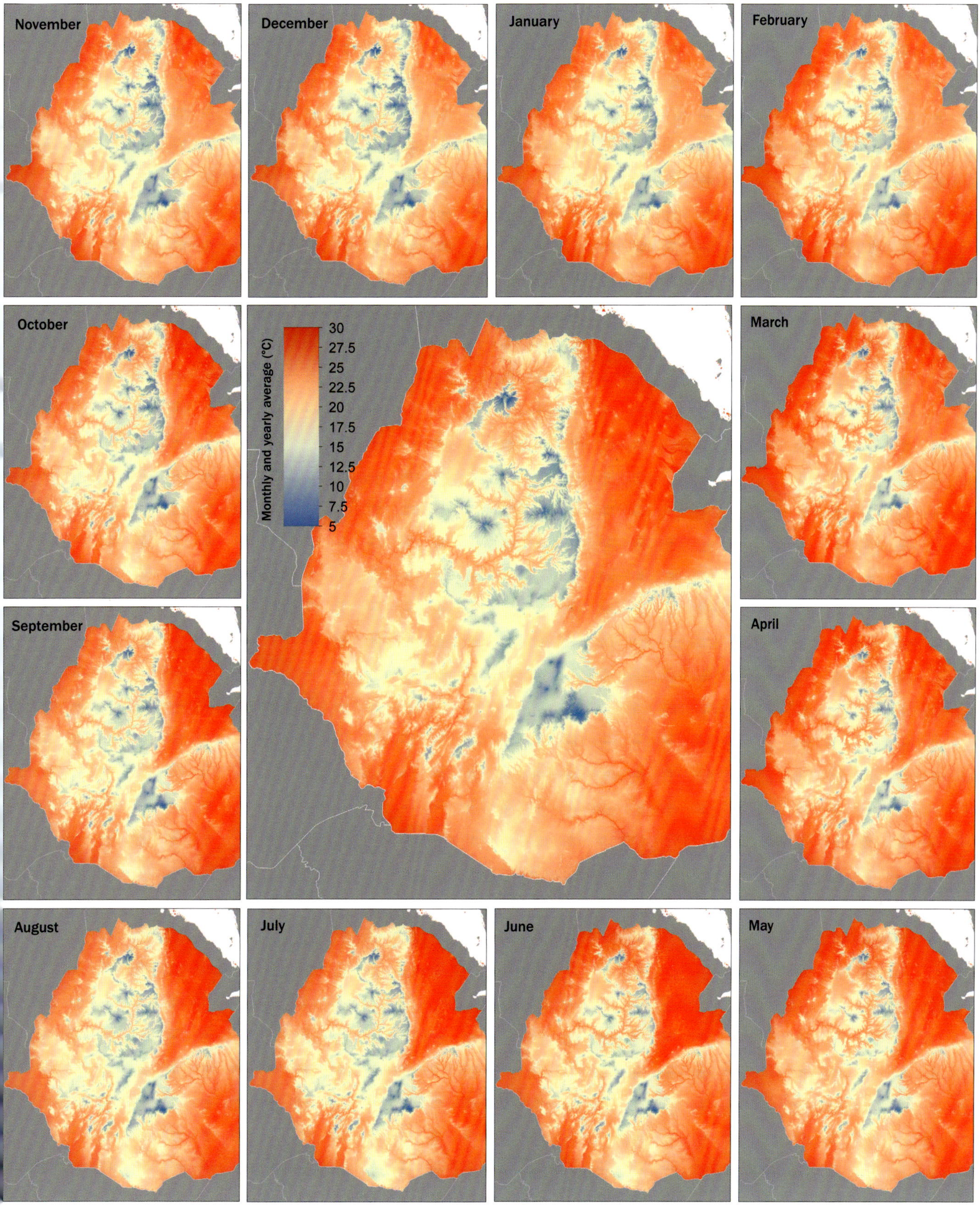

Figure 4. Monthly temperatures from January to December (clockwise) and yearly (central map) average temperatures.

Coffee seedling nursery, with semi-forest coffee in background (Yayu, south-west Ethiopia)
Image: Alan Schaller (courtesy of Union Coffee)

7. Agroecology

Coffee farming and natural vegetation

In Ethiopia, coffee is produced within specific agroecological zones over numerous political divisions (see Table 1, page 50). Most of Ethiopia's coffee is grown in areas that are still covered (Figure 1, page 17), or at one time would have been covered, with humid evergreen forest (Figure 5, page 28). The more precise classification for this type of vegetation is Moist evergreen Afromontane forest (MAF) and Transitional rain forest (TRF)[5]. Currently, about 80% of Ethiopia's coffee comes from these forests or forest-like environments, covering around 20,000 km^2 [10]. The rest is grown in small plots in partial shade or full sun, and covers at least another 4,000 km^2. Humid, evergreen forest is mainly found at 650–2,600 m[5], with coffee-growing mainly confined to elevations of 1,200–2,100 m. These forests are cool-tropical, distinctly seasonal (Chapter 6), and mostly comprise evergreen trees (including wild Arabica coffee) and shrubs, with a high diversity of other plants and associated fauna. In some highland areas, mostly at higher elevations (1,900–2,100 m), coffee farming is undertaken in association with a drier type of vegetation, classified as Dry evergreen Afromontane Forest (DAF)[5], which is dominated by evergreen trees or shrubs, and common native coniferous species (Figure 5, page 28). This type of vegetation is typical of the Harar coffee zone (see page 55), although the natural vegetation has been removed over much of this area.

The association of Arabica coffee and forest is clearly seen at the farm level. Most Ethiopian coffee is grown under the shade of trees (shade or forest coffee), either within forest or forest-like environments, or in farming systems that incorporate specific shade plants — usually indigenous (native) trees or sometimes fruit trees and other crop plants (see below). In some areas coffee is grown with little or no shade (sun coffee), although this probably accounts for less than 20% of total production. Forest (shade) coffee and sun coffee can be considered as two separate production systems but in reality there is a continuum from shade to full sun, and the amount of shade often varies at a single site. Sun coffee farming systems are usually small and mostly fall into the category of coffee gardens, with coffee planted in regular-rows at high density. Sun coffee is usually found at higher elevations (1,700–2,100 m), often near dwellings, and is characteristic of coffee farming in the Harar and Rift coffee zones.

Deforestation

Ethiopia's forest-based coffee farming systems are dependent on the presence of humid forest, particularly for forest coffee (FC) and semi-forest coffee (SFC), as described in Chapter 8. It is clear, then, that deforestation is detrimental for the coffee farming sector. Even though deforestation rates are lower in areas where there is coffee farming[33], it should be recognised that coffee farming can be a significant cause of deforestation and forest degradation.

Measuring deforestation over the long term is problematic as there are various classifications of forest types, and large-scale forest change is difficult to measure successfully, even with satellite technology. Nevertheless, it is clear that there has been significant deforestation in Ethiopia, and this includes the humid (Afromontane) forests[34,35] that are associated with coffee production and wild coffee populations. Based on a calculation of potential humid forest cover and what was present in 2014[10], it is possible to estimate that approximately 63% of this forest type has been lost since pre-agricultural times. Deforestation rates vary according to locality, with some areas having much higher deforestation rates than others[32] (Figure 6, page 29). It is believed that around 70% of the deforestation in Afromontane (humid) forests has been caused by small-holder expansion, intensification, and conversion to agroforestry systems[35].

The coffee suitability niche (climate and forest)

In order to grow successfully, coffee must have the correct climate (Chapter 6), microclimate, and soil. In Ethiopia the microclimate is most strongly influenced by the amount of forest cover/shade, or the lack of it, and whether an area is irrigated or not. As there are very few coffee areas in Ethiopia under irrigation, the addition of extra water to the soil is not a major consideration, at least under current cultivation practices. Tree cover has a major influence on ambient air temperatures, and there may be up to 4°C difference between cooler forest/shaded places and hotter unshaded places within the same locality. Tree cover also influences evapotranspiration and humidity, as well as numerous biological (biotic) interactions, particularly relating to soil, soil moisture, aspects of soil chemistry, and biotic features of the soil. The soil must also have the required physical and chemical characteristics, which are largely influenced by the geology and specifically the underlying rocks (Chapter 3). Other factors influencing the microclimate include coffee cultivation practices, the distance between plants (e.g. planting densities), mulching, terracing and pruning (including stumping). Elevation clearly has a huge influence on coffee suitability, but its main affect is directly related to climate (Chapter 6).

Overall, climate and forest/shade (which influences the climate and microclimate) are the main factors influencing suitability for coffee farming in Ethiopia. In terms of the main climate variables, rainfall, temperature, and seasonality are the most important[10]. Rainfall is especially important, given that suitable average monthly/yearly temperatures for Arabica coffee occur across large parts of the Ethiopian Highlands, and the upper and extreme temperature limits for Arabica coffee cultivation are not seen on a frequent or prolonged basis. A broad overview of coffee suitability, forest cover and forest change across the coffee landscape of Ethiopia is given in Figure 6.

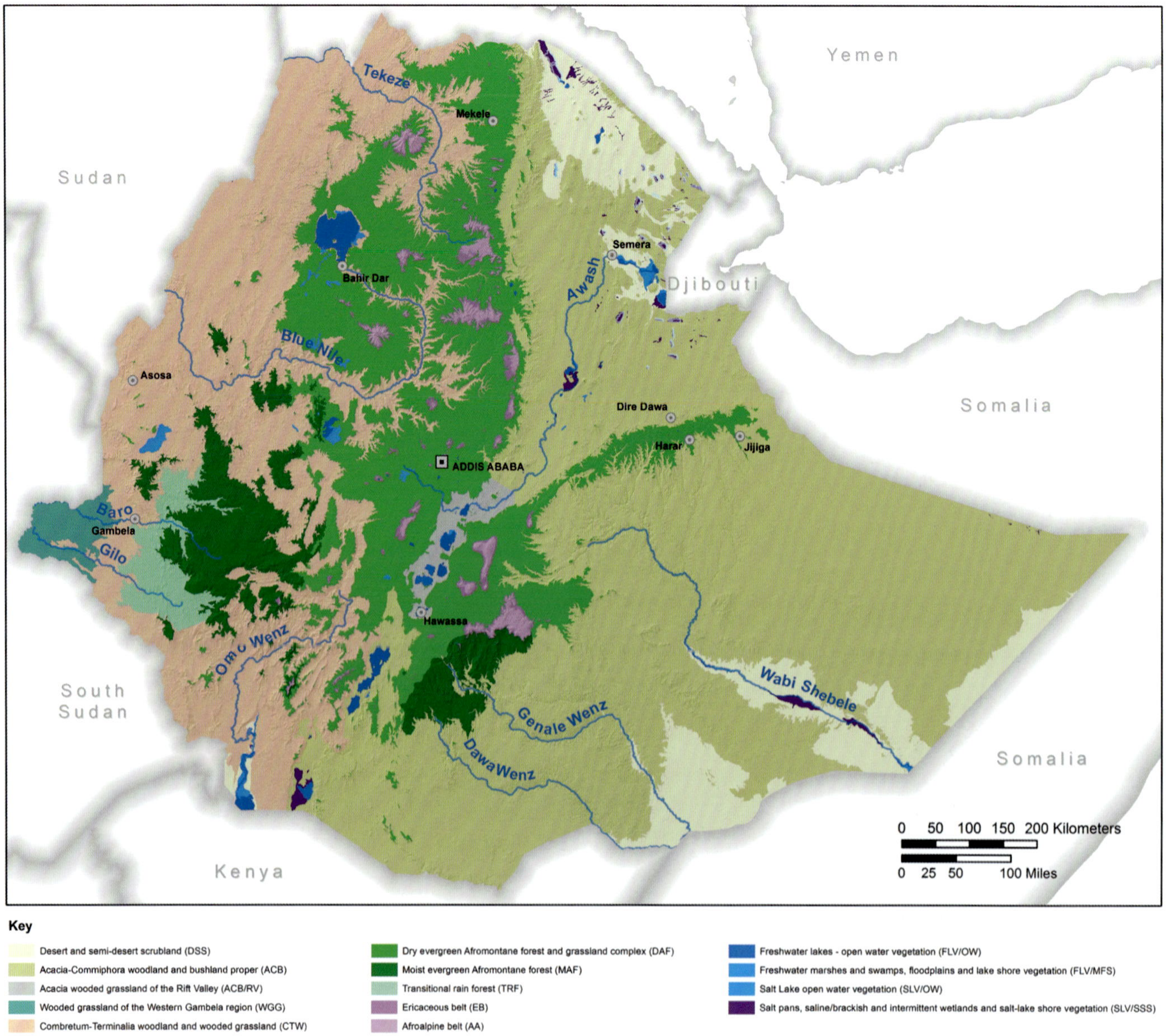

Figure 5. Natural vegetation classification of Ethiopia, from Friis, Demissew & Breugel (2010), *Atlas of the Potential Vegetation of Ethiopia*[5], with major rivers added (see page 5). See key (above) for details of the vegetation classification.

OPPOSITE

Figure 6. Overview of present-day suitability for coffee production in Ethiopia, with modelled coffee suitability (the niche), forest cover and forest change[36]. Map based on climatic variables[10], high-resolution satellite imagery[6], and field observations. See figure legend and text below for further information. Hexagon area is 260 km^2; diameter is 30 km. See Atlas map pages for details.

Forest cover and forest change

Forest change arrows show relative amount of deforestation (c. 2000–2013): polygons without arrows indicate zero deforestation ('static'); small arrows, no or little deforestation; and large arrows ('rapid') a high deforestation rate.

Forest cover hexagons show percentage of forest cover, over a range of 5–70%, from no outline to thick outline.

Coffee suitability (niche)

Potential suitability for coffee-growing or wild coffee forests based on environmental parameters (rainfall, temperature and seasonality)[10].

Unsuitable (no colouration): coffee production is either barely possible or impossible.

Marginal: coffee production is possible but often problematic, with poor and very inconsistent yields.

Fair: coffee-growing is possible and sometimes good, but may require additional inputs (e.g. irrigation, shade or better shade). Yields poor in years with low rainfall.

Good: coffee production is good, with the potential for consistent, high-quality yields. Yield and quality may be negatively influenced in some years (by adverse weather conditions, especially low rainfall and high temperatures).

Excellent: coffee production is good to excellent, with the potential for consistent, high-quality yields. Yields and quality can be negatively influenced by adverse weather conditions, but this is less likely than in other areas.

Coffee gardens

The scale of coffee-growing ranges from a few trees to small plots. Yields may be poor and inconsistent, or good, largely depending on inputs (e.g. irrigation) and/or local conditions (e.g. microclimate, or height of water table).

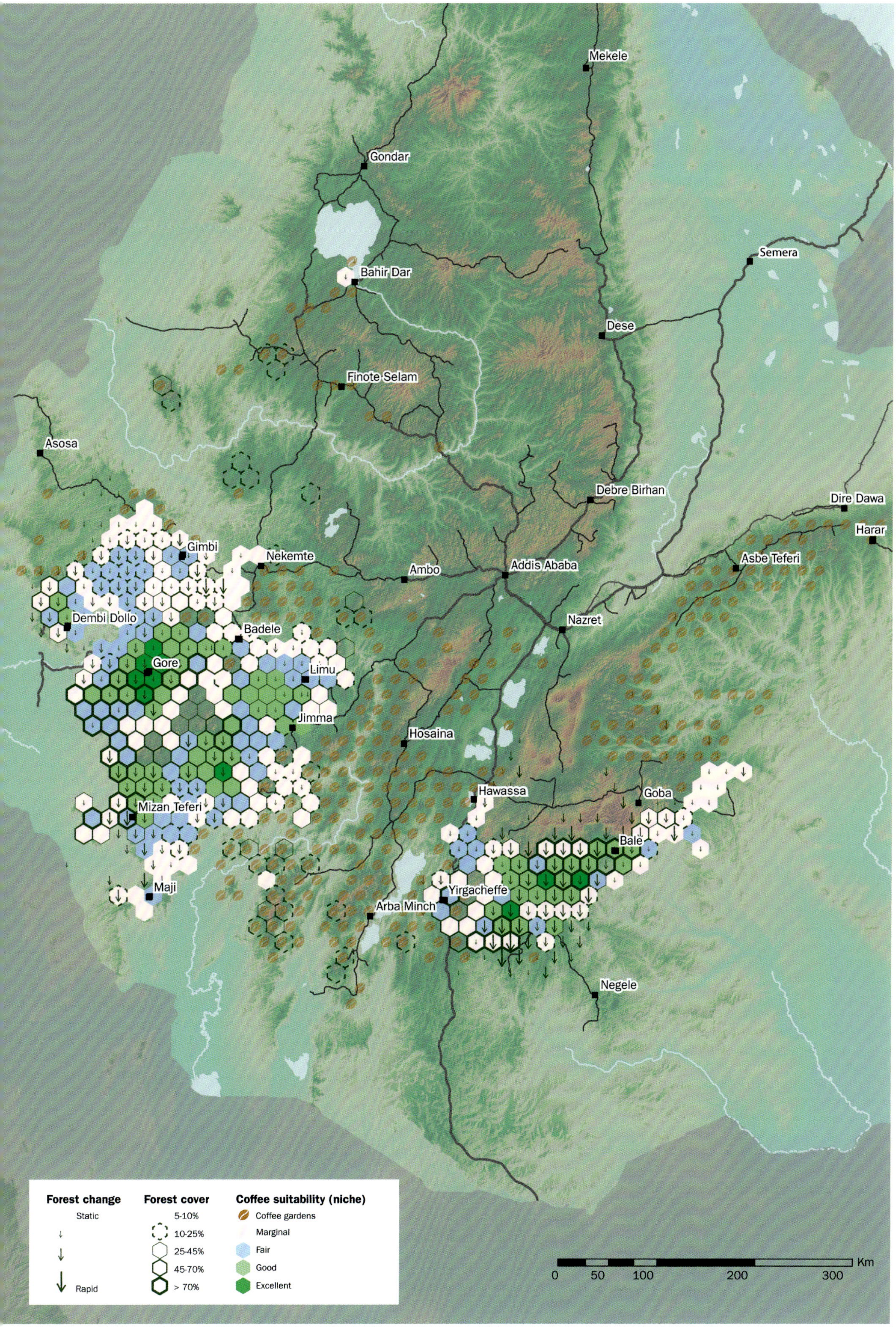

Mekele
Gondar
Bahir Dar
Semera
Dese
Finote Selam
Asosa
Debre Birhan
Dire Dawa
Harar
Gimbi
Nekemte
Ambo
Addis Ababa
Asbe Teferi
Dembi Dollo
Nazret
Badele
Gore
Limu
Jimma
Hosaina
Hawassa
Goba
Mizan Teferi
Bale
Maji
Yirgacheffe
Arba Minch
Negele
Forest change
Static
Rapid
Forest cover
5-10%
10-25%
25-45%
45-70%
> 70%
Coffee suitability (niche)
Coffee gardens
Marginal
Fair
Good
Excellent
Km
0
50
100
200
300

One-year old Arabica coffee plants in a seedling nursery
Image: Alan Schaller (courtesy of Union Coffee)

8. Coffee farming

Coffee farming systems

The characterisation of the forest-based coffee production systems into the categories given below was not straightforward, as there are often no clear or definite boundaries between each system. For example, a forest coffee (FC) system with a high level of management will overlap with semi-forest coffee (SFC), and likewise the SFC with forest garden coffee (FGC). Also, some of the larger and more intensively managed examples of SFC and FGC look more like small, shaded plantations, although generally plantation coffee (PC) is greater than 50 hectares (ha), and commonly more than 100 ha, in extent (see below).

In forest-based production systems, there is usually considerable canopy cover, from around 80% (i.e. that found naturally and in some FC examples), or 50–80% canopy cover as per the management of the farmer (for SFC, some FGC, and agroforestry), with a varying diversity of shade tree species present. Ideal shade levels depend on location, and even within a small area the optimum may differ from farm to farm. For successful shade management, there is the need to achieve a balance between the right growing climate and therefore productivity, and the occurrence of diseases and pests. Generally the optimum is 40–60% shade cover, depending on location.

Managed coffee forests harbour about 60–90% of the plant species that would occur in undisturbed natural forests in the same area, depending on the intensity of management practices and the age of the production system[37]. In more intensively managed forest-based coffee production systems (e.g. FGC, agroforestry, and shaded plantation coffee (PC)), where the canopy/shade cover is 30–50%, the diversity of shade trees (and biodiversity in general) is lower[37]. Generally, forests without coffee production have a higher deforestation risk than forests with coffee production[33]; forests with coffee production have lower biodiversity value than natural, undisturbed forest[33]; and biodiversity decreases with increasing intensification, from FC, through SFC to FGC, but with coffee productivity increasing.

Coffee plantation with native (indigenous) tree species (Gesha, Bench Maji coffee area, south-west Ethiopia) Image: Jenny Williams

Forest Coffee (FC). The FC system uses wild coffee plants, as found in natural forest, with little or no management. This type of production system is not widely practised, especially in undisturbed forest, as productivity is very low (at 50–100 kg/ha, or less). In more disturbed forests, and those with some degree of management, productivity will be higher; in some cases these forests will be in transition to the more intensive semi-forest coffee (SFC) system, as discussed below. FC systems retain the natural forest canopy and all, most, or some of the understory vegetation, and are therefore the best in terms of carbon capture and biodiversity value. Strict FC systems probably account for less than 5% of Ethiopia's total coffee production; all levels of FC production (i.e. strict and slightly managed) are not likely to exceed 10% of total production. Forest coffee is harvested in south-east and southern Ethiopia. Where the land is not owned by the people gathering the forest coffee, it is necessary to acquire a permit from government authorities before harvesting can take place.

Semi-forest Coffee (SFC). The SFC system represents the next stage of intensification from forest coffee (FC), with interventions including the thinning of trees, understory clearance and weeding, and the planting of coffee seedlings. Some of the older SFC areas are derived from the planting or regeneration of native trees species in association with coffee planting, rather than from the modification of natural humid (coffee) forest. In some cases it is difficult to distinguish between SFC and forest garden coffee (FGC). SFC coffee accounts for around 20% of Ethiopia's total coffee production.

Forest Garden Coffee (FGC). This system is the next stage in intensification after semi-forest coffee (SFC), and in most cases the coffee being grown provides a substantial proportion of the farmer's income. FGCs and are typically around 1–2 ha, but sometimes more (e.g. 5 ha or more). The FGC coffee farming area is predominately dedicated to coffee, but there can be other crops growing nearby and sometimes intercropping is used. The canopy (shade) cover in FCGs is less than in FC and SFC, and there is an increase in farming activities, including more regular and denser planting of coffee trees, pruning and removal of shade trees, regular weeding, and pruning (stumping). FGCs are by far the most common of the smaller-scale forest-based coffee farming systems, and the most productive (see section below: Productivity). FGC makes up around 30% (and possibly more) of the total coffee production of Ethiopia.

Coffee Farms (plots). This system of classification is similar to forest garden coffee (FGC), but provides a more general term to include all farms, both forest-based (including agroforestry) and otherwise, including unshaded and lightly shaded systems.

Agroforestry. Agroforestry is similar to the forest garden (FGC) system, and is sometimes included within the FGC system category. However, special mention is required for agroforestry because of the specific characteristics associated with

Coffee garden in full sun (near Hirna, West Hararge coffee area, central-eastern Ethiopia) Image: Aaron Davis

this production system. Agroforestry systems are generally smaller and more intensive than FGC in terms of the number and diversity of crops being growing. Key crops in addition to Arabica coffee, include enset (*Ensete ventricosum*), maize (*Zea mays*), khat (*Catha edulis*), and yams (*Dioscorea* species). As the name suggests, this is a forest-based system, that mostly uses indigenous trees to provide the shade for coffee and other crops (see section below: Shade plants). Agroforestry farms are typically family run, of between 0.25 and 1 ha, and provide food for the family and income from the sale of farming produce. Coffee is grown for home or local sale, and on larger agroforestry smallholdings for export. In the western Sidamo area, for example, around and near the towns of Aposto, Dilla, Wenago, Yirgacheffe, coffee is almost exclusively produced within an intensively managed agroforestry system.

Garden Coffee (GC) or Coffee Gardens. Garden coffee (GC) is often planted close to dwellings, with the number of trees planted ranging from just a few plants up to around 100. GC is usually part of a mixed cropping system and provide coffee for the household, the local market, and where productivity is good, the crop may be collected from many coffee gardens for export. Coffee gardens are planted in either full sun or in semi-shade, with the shade provided by native or introduced tree species. GC is not part of the forest-based production system, and contributes around 20% of the total coffee production of Ethiopia.

Plantation Coffee (PC) or Coffee Plantations. Coffee plantations are large and intensively managed systems, either privately or state owned. There is no fixed size limit for a plantation but generally they are more than 50 ha (0.5 km^2). Very large commercial plantations (e.g. over 500 ha and up to 10,000 ha) are uncommon. Plantation coffee is the most intensive type of coffee farming in Ethiopia, and comprises uniformly planted coffee trees in neat rows, usually with good access for cultivation activities, farm machinery and harvesting. Depending on the individual plantation, various intensive inputs are made, including organic and inorganic fertilisers, pesticides and herbicides, or some form of non-chemical pest and disease management. On better plantations integrated weed and pest management systems are in place. The coffee trees may be pruned, or stumped, on a regular basis. Soil mulching is practiced on some plantations, including the use of coffee hulling waste. In some plantations, irrigation is used, but so far this tends to be of lower technical sophistication compared to other producing countries (e.g. in South America and Asia). Ethiopian coffee plantations usually incorporate shade (tree) cover, but in many cases the density of shade-tree planting (and therefore the level of shade) is often less than that in forest-based coffee systems, and is much more precisely managed. Other farming activities include the use of soil erosion control in the form of water diversion ditches or bunds (low earth/soil walls), or low-lying hedges of soil/water retaining plants, such as vetiver grass (*Chrysopogon zizanioides*).

The genetic diversity of the planted coffee stock used in plantations is usually less than that in traditional coffee production systems (see above), with selected cultivars and selected lines used. In rare cases on lowland plantations (e.g. around or below 1,000 m), robusta coffee (*Coffea canephora*) may be planted. Harvesting is still done by hand, unlike many large plantations in South America, and especially in Brazil, where harvesting is achieved using specialised farm machinery.

The coffee plantations at Bebeka, in the Bench Maji coffee area (Figure 7, page 49) near the town of Mizan Teferi (Atlas map page 9), is currently the largest in Ethiopia at around 10,000 ha. A sister plantation of some 8,000 ha, the Limmu Coffee Farm, is located in the Limu part of the Jimma-Limu coffee area. Previously government-owned, these are now in the private sector. Perhaps the oldest coffee plantation is in Harar (Arsi) near the small town of Gololcha, which dates back to the early twentieth century.

PC is the most productive coffee system in Ethiopia, as illustrated in the section below on productivity. Coffee plantations are not common in Ethiopia at present, although due to higher productivity, and in some cases higher quality,

Semi-forest coffee (SFC), recently planted near Metu, Illubabor coffee area, south-west Ethiopia Image: Aaron Davis

economies of scale, and good marketing, it is possible that they will increase over the coming decades. At the present time PC contributes 5–10% of total coffee production for Ethiopia.

Shade plants

Forest and shade-based coffee production systems use native tree species. The diversity, i.e. the number of different species used, is largely related to the coffee production system in place. With increased intensification and management (e.g. from FC, through SFC, to FGC), the number of tree species declines. Certain tree species are encouraged, sometimes the more popular types are planted, and in some cases certain trees species are removed if they are considered to be disadvantageous to coffee tree health and productivity. Popular and common indigenous shade species include *Acacia abyssinica* (Leguminosae), various species of *Albizia* (Leguminosae), *Bersama abyssinica* (Melianthaceae), *Celtis africana* (Ulmaceae), *Croton macrostachyus* (Euphorbiaceae), *Cordia africana* (Boraginaceae), *Ekebergia capensis* (Meliaceae), *Erythrina abyssinica* (Leguminosae), various species of *Ficus* (Moraceae), the very popular shade tree *Millettia ferruginea* (Leguminosae), *Olea capensis* subsp. *welwitschii* (Oleaceae) *Polyscias fulva* (Araliaceae) and *Syzygium guineense* (Myrtaceae)[38]. Not only do these trees provide shade for the coffee plants, but they also protect coffee flowers from damage by heavy rain, provide a nectar source for coffee pollinating insects (bees), may fix nitrogen (i.e. the species of Leguminosae), produce mulch (as leaf litter), and assist in soil and water conservation.

In smaller scale, intensively cultivated coffee production systems, such as garden coffee (GC), agroforestry, and some types of forest garden coffee (FGC), non-indigenous fruit trees, such as mango (*Mangifera indica*; Anacardiaceae) and avocado (*Persea americana*; Lauraceae), are used to provide shade. However, both of these trees produce dense shade and take considerable moisture from the soil, so the density of planting in these coffee production systems is low. In areas that grow the staple crop enset (*Ensete ventricosum*; Musaceae), particularly the agroforestry systems in Sidamo (e.g. in Yirgacheffe), this banana-like plant is often very successfully used to provide shade for coffee.

Non-indigenous shade species are mainly used in the initial establishment of shade, and to provide shade and wind protection for young plants, particularly in coffee plantations. These include: the silky oak (*Grevillea robusta*; Proteaceae), a medium sized tree; the castor oil plant (*Ricinus communis*; Euphorbiaceae), a small to medium-sized shrub; and the pigeon pea (*Cajanus cajan*; Leguminosae), a medium-sized, short-lived herbaceous plant.

Productivity

Conversions (to clean coffee)

The standard conversion of ripe, fresh coffee fruits (cherry) to clean coffee (unroasted/green beans) is 5:1 (a multiple of 0.2), which means that 5 kg of fresh cherry would yield 1 kg of clean coffee, after processing and hulling. Cherry to clean coffee conversion rates can be lower, for example 6:1. These rates do not include cherry that is discarded during processing, but rather the loss in weight due to drying and the removal of the pulp and parchment. The conversion of dried cherry (produced by dry (natural) processing) to clean coffee is 2:1 (a multiple of 0.5); thus, a 60 kg sack of dried cherry would convert (after hulling) to 30 kg of clean coffee. For parchment coffee, the conversion is 1.25:1 (a multiple of 0.8), and so a 60 kg sack of parchment would convert to 48 kg of clean coffee. Working from clean coffee, backwards, a 60 kg sack of green coffee approximates to 300 kg of fresh cherry, 120 kg dried cherry, and 75 kg of parchment. When green coffee is roasted it loses approximately 20% of its weight, mostly due to the removal of moisture, so that a 60 kg sack of green coffee would yield around 48 kg of roasted coffee.

The weight unit 'quintal' is frequently used and quoted in Ethiopia, for the yield of coffee and other seed crops. A quintal is 100 base units, i.e. one quintal is equal to 100 kg.

Production figures

Ethiopia is Africa's largest and the world's fifth largest coffee exporter[39]. In 2016 it exported three million (60 kg) sacks of (green, unroasted) Arabica coffee, equivalent to 180 million kg (180,060 metric tonnes). These figures do not, of course, include the coffee that is consumed in Ethiopia, which is estimated to be about the same amount, or slightly more, than the coffee exported. Since 1990 the amount of Arabica coffee exported by Ethiopia has almost tripled[39]. Despite these impressive figures, productivity per unit area (kg per hectare (ha)) is low when compared to that in many other Arabica coffee-producing countries, particularly in Central and South America.

The yields quoted below refer to clean (green) coffee. Estimated and calculated values for the different production systems include: forest coffee (FC) 50–200 kg/ha; forest garden coffee (FGC)/coffee farms (plots) 300–500 kg/ha, or more productive examples, 600–750 kg/ha; and for well-managed plantation coffee (PC) in areas of high climatic suitability, 600–1,000 kg/ha, or perhaps more. The average coffee yield across the country is often stated to be around 700 kg/ha, although this may be on the high side, and other sources suggest that the average yield is between 300–400 kg/ha. One must also bear in mind the factor of biennial bearing, which often results in marked differences between one harvest and the next (see section: The biennial cycle). The figure of 300–400 kg/ha seems reasonable based on farmer feedback. For example, in Illubabor where a good harvest might reach up to 600–800 kg/ha in an 'on' year in the biennial harvest cycle, and 100–250 kg/ha for a poor harvest (in an 'off' year). Ethiopian production figures are much lower than the most productive Arabica coffee-producing countries, for example Brazil and Costa Rica, where farms consistently produce around 1,500 kg/ha (or more) and 780–1,000 kg/ha, respectively.

The reasons for low productivity in Ethiopia are many and various, including: low planting densities due to the space required for shade trees; climatic constraints (e.g. the long dry season, and general low rainfall); ageing stock (e.g. lack of pruning/stumping, or replacement by seedlings); minimal or lack of fertiliser use (organic or inorganic); diseases and pests; and soil health and soil erosion. Market demand and coffee prices influence nearly all of the above factors; farmers generally produce more coffee and of a higher quality when prices are high and stable.

It can be useful to understand the amount of clean coffee produced for each coffee tree. A figure of 5 kg fresh cherry (resulting in 1 kg of clean coffee) per tree is often quoted by farmers and other sources, based on harvesting figures and experience. This is the upper end of the range for Ethiopia, with 3–4 kg of fresh cherry (600–800 g clean coffee) being more realistic, and around 1 kg of cherry (200 g clean coffee) at the lower end of the range. Arabica coffee produces between 1,000 and 3,000 fruits per tree, and so if a single cherry weighs around 1.8 g, that would give a figure of 1.8–5.4 kg of fresh cherry, and 0.36–1 kg of clean coffee, per tree.

The cycles of coffee production

The yearly cycle

The flowering, fruiting and growing cycle of Arabica coffee, is controlled by climate (Chapter 6), which is closely linked to location, and the inherent physiological characteristics of the plant itself (Chapter 5). The dormancy of the flower buds is broken by the first rains of the wet season, which usually start in March or April, depending on the locality. The flowers are short-lived and fragile, and the young fruits start to form soon after fertilisation. In the first 6–8 weeks the fruits remain small and round and are referred to as 'pin-heads'. Over the next 6–7 months the fruits slowly increase in size, and from then onwards, the fruit starts to become fleshy and watery, turning from green, through to yellow or green-yellow, and then finally red at maturity. The speed of development and maturation depends on the season's weather, local climate and elevation (higher elevations being later). Most of the fruit development stages coincide with the early period of the wet season, as does the initiation and growth of new leaves and the growth of the leaf-bearing shoots and the main stem. Fruit ripening coincides with declining rainfall, as the yearly cycle approaches the dry season. In total, fruit development, from fertilisation to full maturity, takes about 9 months. Further details of fruit development are given in Chapter 5. The Ethiopian coffee harvest period usually coincides with the early to middle part of the dry season, between September and November, but it can start as early as August and as a late as February, and sometimes later. There is a lot of regional variation in harvesting time, including: November to early January in most of the South West coffee zone; December to January in (Kelem) Wellega; September to December in the western part of the Sidamo coffee area (South East coffee zone); December to February in Arsi (western part of the Harar coffee zone); January to March in northern locations in the North coffee zone (e.g. on the Zege Peninsula); and October to December (or sometimes as early as August or September) for irrigated farms in the North coffee zone (e.g. at Finote Selam). Whilst

Newly germinated coffee seedlings, or 'soldiers' Image: Aaron Davis

being fairly consistent, regional harvest times are influenced by yearly weather patterns and elevation, with higher elevations being later, and *vice versa*.

After the harvest comes the resting period for coffee, which coincides with the remainder of the dry season. The dry season, which can last for 3–4, or even 5 months, may see light rains or very little rain at all (Chapter 6). During this dry period the flower buds develop, but vegetative (leaf, stem and woody) growth is negligible. There is a considerable amount of leaf-fall in the dry season and the leaves often wilt to varying degrees; both leaf-fall and wilting increasing in hotter, drier conditions. Generally, the coffee plants look tired and unhealthy during, and particularly towards the end of, the dry season. With the onset of the rains the yearly cycle recommences: with flowering, the flushing of the new season's leaves, followed by the growth of the shoots and the main stem, and fruit development.

The biennial cycle

Biennial bearing is a regular feature of Arabica coffee production in Ethiopia and East Africa, and indeed throughout the coffee-growing world. In biennial bearing a good or excellent (bumper) crop is followed by a poor, low-yielding crop in the next year's harvest. For this reason biennial bearing is referred to in terms of 'on' and 'off' years. This phenomenon occurs independently of weather and climate, so that even when climatic conditions are favourable for a good crop, the yield can be poor. There can be a fivefold (× 5) difference between on and off years, although in extreme cases there can be a tenfold (× 10) difference in Ethiopia, as in other areas of East Africa[40]. The biennial cropping rhythm is due to the allocation of the plants resources: in the productive 'on' year, the coffee tree puts its resources into producing the crop at the expense of vegetative growth (stem growth and particularly shoot development and flower bud production); in the 'off' year, the plant must make up for this vegetative shortfall, at the expense of flower and fruit production. Biennial bearing can be reduced or almost eliminated using various means, including pruning, the addition of fertiliser, controlled irrigation and the use of selected vigorous clones or cultivars.

Seed and seedling production

Coffee plants can be propagated by various means, such as cuttings, tissue culture, and grafting, but in Ethiopia the production of new planting stock is almost exclusively undertaken via the sowing of seed. To do this, ripe fruits are removed from the coffee plant and the pulpy parts are removed (i.e. leaving the parchment) either by hand or with a pulping machine, with or without the use of water. Afterwards the seeds are treated in various ways, for example they are often coated with wood ash or powdered charcoal. If correctly prepared and stored, including drying to a level (15–20% moisture) that inhibits pathogens (bacteria and fungi) and yet does not affect the viability of the embryo, the seed may be kept for several months. As seed storage time increases, germination success decreases, and germination time increases, especially after about 3 months of storage, although viability is also strongly influenced by storage temperature and seed moisture. The seeds can also be sown straight away, following preparation. For fresh seed, the germination time is about 7–10 weeks, although in good conditions this could be as short as 4–5 weeks.

The seeds are either sown directly into the ground, into seedbeds with prepared soil, or into polythene pots (polybags), in specially prepared shade beds, mostly fabricated from easily obtainable plant materials such as coppiced tree shoots, grass, and large leaves. The shade beds are either irrigated or the plants are watered with a hose or watering can.

Shortly after germination the seed is pushed above the soil on the young stem, at which stage the seedling is referred to as a 'soldier'. After 3–4 weeks the seed leaves (cotyledons), which are approximately ear-shaped and have wavy margins, force off the old seed. Soon afterwards the true leaves appear and then further leaf pairs develop, at which stage the cotyledons fall off. The seedlings are left to develop in the shade beds for up to one year, by which time they are around 15 cm tall and easily transportable. Seedlings in polybags tend to have better survival and establishment rates than bare-rooted seedlings, although some farmers prefer the latter because a larger number of seedlings can be transported and planted more easily. Planting is usually undertaken once the wet season is well underway, mostly from May through to August.

Planting densities

Planting distance and spacing vary widely in Ethiopia, depending on personal preference, tradition, local conditions and climate, the soil and other growing conditions. Commonly used planting distances are 1.5–2 m between each plant, and 2.5–3 m between rows. At a distance of 1.5 m between trees and 2.5 m between rows, there would be 2,666 coffee trees per hectare (ha); and with 2 m between trees and 3 m between rows, there would be 1,666 coffee trees per ha. The number of trees per ha can be many more, especially on farms that are irrigated, with figures quoted of around 5,000 per ha, or even higher. For most Ethiopian farms (i.e. non-irrigated), the number of trees per hectare is usually 1,000–2,000. In areas where the climate is not ideally suited to coffee, the density can be as low as 500 trees per ha. It is often the case that some of the coffee trees are either missing (e.g. they are dead or badly damaged) or are being replaced (by seedlings). Moreover, in Ethiopian shade-based production systems, a considerable proportion of each hectare is taken up by the space required for shade trees.

General cultivation

Pruning and stumping

Pruning in Ethiopia is mainly restricted to stumping, a method of rejuvenating ageing coffee trees by cutting them down, almost to ground level. Ideally this should be undertaken every 8–10 years, but in Ethiopia it is often delayed to around 15 years, or more. Stumping is undertaken during the dry season, after harvesting and before the onset of the wet season, for example from December to February, although the timing varies. The coffee trees are cut down so that only a short length (20–30 cm) or longer (c. 50 cm; known as clean stumping) of main stem (trunk) remains; the remaining stem is referred to as the 'stool' in English. The cut through the main stem is made at an angle of around 45°, preferably with a sharp saw, so that water does not collect on the stump and aid infection by pathogens. Sometimes, if a multiple-stemmed tree is being stumped, one of the stems (trunks) is left in place, until the new stems have grown and established (see below). After 3–6 months numerous new shoots develop from the stool, but before they grow too much, the vast majority are cut away (from the base) to leave three to five, well-spaced healthy shoots. One year after stumping, the new shoots may be further reduced in number (e.g. to three) for a multiple-stemmed tree, or reduced to a single main stem, for a single-stemmed tree. If one of the original stems was left from the old tree, it is also removed at this stage. It is not in the farmer's interest to stump all of his trees at once, and so a selected number (or area) are stumped, so that an income can be maintained from the non-stumped trees. Two years after stumping, the coffee trees will come back into productivity and after 3–4 years they should significantly outperform the non-stumped trees, in terms of coffee yield[41]. Research undertaken in Ethiopia shows that other methods of pruning Arabica coffee can increase yields in comparison to the traditional stumping method[42]. Despite the yield benefits of stumping, many Ethiopian farmers are reluctant to stump their trees, probably because of the short-term loss in income, even if the trees are in low productivity due to old age.

Weeding

Weeding is an important part of post-planting aftercare and the general maintenance of coffee farming land. The coffee planting area is cleared of weeds before the planting of new stock, and thereafter two to three times a year, although in drier parts of Ethiopia, only one round of weeding is required. Weeding is undertaken by slashing with a machete, or a similar tool, and the weeds are either left where they are cut, tidied up to the edges of the field, or removed. Herbicides are almost never used, except perhaps in large, intensively managed plantations.

Control of pests and diseases

As in many other coffee-producing countries, diseases and pests can account for a substantial reduction in coffee productivity in Ethiopia. Diseases are generally more problematic than pests. The major diseases of coffee in Ethiopia are coffee berry disease (CBD) and coffee wilt disease (CWD).

CBD is caused by the fungal pathogen *Colletotrichum kahawae*. The CBD pathogen attacks the leaves, flowers and fruits, but it is on the young, developing fruits (1–4 months old) that the symptoms are most readily seen. Initially the fruits have one to several brown and sunken lesions, which develop and then cover the whole fruit, turning it dark brown or black, and wrinkled. The stem on which the affected fruits are attached

Fruiting shoot showing symptoms of coffee berry disease (CBD) Image: Aaron Davis

may also become blackened. The fruits may also have scab-like lesions, which develop on the fruit as it ripens. Crop losses per unit area can be as low as 10–20%, or much higher, up to and over 80% in extreme cases. As CBD appears to be linked to high rainfall and high humidity, the management of the farming microclimate is important, and better air circulation (e.g. increasing the planting distances) can help to reduce CBD. Where available and affordable, copper-based fungicides are sometimes used, and these can be effective. The planting of CBD-resistant stock, including that selected from wild Arabica stock, has been effective in some parts of Ethiopia.

Coffee Wilt Disease (CWD) is caused by the fungal vascular wilt pathogen *Gibberella xylarioides*. The symptoms are often first seen on new and developing leaves, particularly in young plants. In the early stages, the leaf tissue close to the midrib and main veins becomes brown and necrotic (dead or dying), and a short while afterwards the whole leaf dries, shrivels and falls from the plant. In mature plants the leaves turn yellowish, wilt and curl, and fall to the ground. In this way even mature plants can become rapidly defoliated, with blackened (necrotic) shoots and fruits (if present). In order to distinguish this pathogen from other diseases, the bark at the base of the main stem should be scraped off: if there is a blue-black staining, particularly at the interface of bark and wood, CWD is the most likely cause[43]. Management of CWD in Ethiopia is undertaken by cutting down infected trees, and sometimes all of the trees surrounding the infected plants, and burning them. Planting CWD-resistant Arabica cultivars can be an effective course of management. Infection rates of planting stock can be reduced by extreme vigilance, sanitary cultivation practices, and the planting of disease-free and disease-resistant plants.

Coffee leaf rust (CLR), *Hemileia vestarix*, is a serious fungal pathogen of Arabica coffee, and the cause of recent devastating crop losses in Central and South America. The most obvious symptom is the presence of pale yellow, mould-like, powdery spots on the under-surface of the leaf. CLR is common on both wild and cultivated Arabica plants throughout Ethiopia, although the level of infection is usually low and with little or no effect on the plant or its productivity. It is not understood why this is the case, but the relatively cool and dry climate of Ethiopia compared to those of many other more tropical Arabica-producing countries (which are hotter and wetter) may be a significant contributing factor. In addition, higher levels of genetic diversity in indigenous Arabica coffee, and the more natural production systems used in Ethiopia, may also contribute to lessening the negative consequences of CLR. There are other minor fungal and bacterial diseases of Arabica coffee in Ethiopia, but generally they do not present any serious problems.

Under-surface of coffee leaves with symptoms of coffee leaf rust (CLR)

Image: Aaron Davis

The main coffee pests in Ethiopia are leaf miner, coffee berry borer (CBB), stem borers, scale insects, and various leaf-cutting organisms. Overall, their influence is minor compared to CBD and CWD. Coffee berry borer (*Hypothenemus hampei*) is a small, brown-black, beetle, 1.2–1.8 mm long; the adults bore into the fruits and the larvae eat the developing coffee beans (the endosperm of the seed), which can lead to large reductions in yield. CBB has only recently been found in Ethiopia, although it could become a major pest[44].

Irrigation

The vast majority of Ethiopian coffee is rain-fed, that is, it relies wholly on the seasonal rainfall. Irrigation is mostly confined to places where water is easily available and can be diverted to the farm using simple, practical means, for example diversion from rivers using trenches. Irrigation is mainly practiced in Amhara and Benishangul Gumuz coffee areas (North coffee zone), and in the north-eastern part of the Oromia Region in the Harar coffee zone (Figure 7, page 49; Table 1, page 50). In the Amhara Region, the town of Finote Selam and places close to this town, are particularly noted for the irrigation of coffee using rivers and streams.

Feeding (use of fertiliser)

The use of fertilisers is generally minimal compared to that in many other coffee-producing countries, with the exception of the larger and better managed plantations. In most cases, feeding is either zero or minimal, and if undertaken is largely restricted to the use of composted animal waste products and composted farm and household waste, in other words, by organic methods. The application of fertilizer, and particularly nitrogen, can significantly increase coffee yields.

Planting stock (landraces, farmer's lines, heirloom varieties and cultivars)

All Ethiopian coffee planting stock was originally derived from the country's wild populations of Arabica coffee. Even today, much of the seedling-derived stock is sourced directly from the forest. A large proportion, however, has been selected and developed by Ethiopian coffee farmers and agriculturalists in order to capture specific traits, such as higher productivity, ease of cultivation, and resistance to certain diseases. These selections are referred to as 'landraces', 'farmer's lines', 'farmer's varieties' or 'heirloom varieties', although the last of these terms is often used as a general term to denote a

Under the canopy of mature coffee trees (Yayu, Illubabor) Image: Emily Garthwaite (courtesy of Union Coffee)

mixture of coffees gathered from several or many farms in a specific locality. There are many hundreds of Arabica coffee landraces in Ethiopia, most of which are specific to their respective place of selection. It is often said that more than one landrace is planted on a single farm, each used for a specific purpose, but this is not the case on smaller farms. Characteristics of tree shape, leaf colour, weight, shape and size of fruits and beans, as well as roasting features, are used to distinguish landraces and may feature in the naming of them. A useful review of landraces is given by Gole *et al.* (2016)[38], including a list of traditional landraces recorded in different coffee-growing localities. The evaluation of Ethiopia's Arabica landraces needs further careful research, particularly at the molecular (DNA) level.

Cultivar is the preferred formal term for cultivated varieties, the term being composed from parts of these two words, i.e. 'culti' and 'var', to give cultivar. The term variety should be reserved for formal, botanical (naturally occurring) varieties. The term variety is, however, widely used in the coffee sector, as is 'varietal', which has probably been adopted from the wine sector. In the older literature, certain origin-specific Arabica variants are often mentioned. For example, 12 'coffee types' are listed by Sylvain (1958)[26]: Ennarea (or Ennaria), from the modern-day Jimma-Limu coffee area; Kaffa; Agaro; Cioiccie; Irgalem, an earlier name for Yirgalem; Dilla; Arba Gougou; Harar; Zeghie (Zege); Loulo; Wolkitte, now called Welkite; and Wollamo (an area situated around the town of Soddo). Nekemti (or Lekemti, and similar name variants) was another commonly used name, for the coffee originating from areas close to (but mostly to the west of) the town of Nekemte in East Wellega. A few of the cultivars listed above make use of the colour of the new leaves and leaf tips (bronze vs. green), which has been given prominence for recognising two major coffee types being grown outside Ethiopia, i.e. 'Bourbon' and 'typica' (see Chapter 5). Although the numerous names listed above appear in coffee research literature, many of them have little practical application today. Nevertheless, provenance still forms the basis of the modern-day classification for Ethiopian Arabica coffee diversity (Chapter 10), and four names (Harar, Limu, Sidama/Yirgacheffe and Wellega) provide the group names for the 37 coffee cultivars distributed by the Jimma Research Center (JRC)[45]. These cultivars were developed to be high-yielding and disease resistant (to CBD)[46], and several million of have been distributed to farmers across Ethiopia.

Harvested coffee fruits (fresh cherry)
Image: Alan Schaller (courtesy of Union Coffee)

9. Coffee harvesting and processing

Factors influencing quality and taste

The quality of unroasted (green) coffee depends on many factors, including those mostly outside human influence, such as climate, geology/soil, and genetics (except in the case of plant selection); and those directly influenced by the primary producer, such as farming practices, harvesting and post-harvest processing. Climate is a key factor, as temperature and soil water availability (mainly due to rainfall) have a direct bearing on the growth of the plant, flowering, and in particular, the development of the fruit. Elevation has a direct bearing on the local climate, and other factors such as light levels and the quality of light (e.g. ultra violet (UV) light). Soils can influence the crop in various ways, including the way the soil retains and makes water available to the plant (physical features), and by virtue of its chemical (mineral) composition. Arabica coffee variants (e.g. cultivars and landraces; see Chapter 8) have specific quality and flavour characteristics, including body ('fullness' or overall feel (light, medium, or full) of coffee in the mouth), acidity and flavour notes. Many of these characteristics are encoded within the DNA of the plant but are influenced in various ways by the environment. Thus, there is a close interplay between the genetic makeup of the plant and its environment. Importantly, the environment influences the chemistry of the coffee bean (seed) and its external fleshy parts (the pulp), and this has a direct bearing on quality and taste, even from year-to-year.

Included in the narrative of Chapter 10 (The coffee areas) are overviews and examples of the various flavour profiles for different coffee areas of Ethiopia, which were collated from commercial tasting (cupping) descriptions and dedicated cupping sessions undertaken for the Atlas. It is clear from this information that 'origin' is a very real phenomenon in Ethiopia, in terms of linking a location or specific area to sensory characteristics (e.g. taste). It is also important to remember that for Arabica coffee in Ethiopia both climate and genetic identity[47,48] are tightly linked to location.

Whilst it may be difficult to pinpoint a flavour profile to any specific coffee area, the differences between (and even identification of) many of the coffee areas can be made by experienced coffee tasters (cuppers). There is an unmistakable distinction in flavour profiles between areas east and west of the Rift Valley (see Table 1 for division of coffee areas): on the eastern side of the Rift Valley (e.g. Sidamo, Bale, Harar) the coffees typically demonstrate bolder flavours, with more pronounced fruit notes, compared to those from the western side of the Rift Valley (e.g. Limu, Kaffa, Wellega). Overall, Ethiopian Arabica coffees stand out from domesticated cultivars grown outside Ethiopia due their more complex flavour profiles.

It should also be made very clear that in addition to the above factors, harvesting and post-harvest processing have substantial influences on the taste and quality of Arabica coffee, as outlined below.

Harvesting

All Ethiopian coffees are harvested by hand-picking, which is a very labour-intensive and skilled process. The colour of the fruits (cherry) at harvesting, which equates to the state of maturation, is critical to cup quality. Varying degrees of red represent the fully mature stage of the fruit, but an even, medium to rich red is mostly sought by well-trained and conscientious pickers. Green, immature fruits, often referred to as 'quakers', are underdeveloped and lack the high sugar content and acidity level found in the mature fruit (and bean), giving coffee a 'dry' mouth-feel (astringency) and a grassy or straw-like taste. In addition, the beans are usually light in colour and weight and do not roast properly. Over-mature fruits are deep, dark red or purple-red, and should be avoided as they can give the coffee a sour (hence the term 'sours') or fermented taste; in the worst cases, the beans produce foul-tasting coffee, and these are commonly referred to as 'stinkers'. If processed correctly, ripe fruit yields sweet, clean coffee, without the defects mentioned above.

Picking is either done cherry-by-cherry, or by stripping a shoot bearing a large proportion of ripe cherry in a single action; the former method improves coffee quality. The coffee fruits do not usually ripen synchronously, with trees bearing green (unripe) and red (ripe) fruits at the same time. Pickers often have to return to the same trees multiple times, and up to five times, to harvest all the mature fruits. This considerably extends the harvesting season for any one farm. Sometimes

Pre-sorting coffee fruits (cherry) prior to pulping and washing

Image: Jenny Williams

Sun-drying coffee fruits (cherry) on a concrete patio Image: Jenny Williams

there is an early 'fly-crop', which is either picked or not picked; picked fly-crops may form part of the overall harvest, or may be picked early and discarded with a view to improving the main harvest. During harvesting the cherry is placed in a small container or basket and then put together to fill a standard- sized sack for transport to either the drying (for naturals) or washing (for washed coffees) station (see below). The coffee bags containing the cherry are weighed, either at the farm/co-operative or the processing centre, so that the weight can be ascertained and payment made to the picker(s), or to middlemen (*akrabi*) who trade between pickers and other buyers. The interval from picking to processing is critical, and in most cases the fresh cherry must reach the processing area as soon as possible and at least within a day of harvesting, otherwise the cherry will start to ferment and quality will be negatively influenced.

Initial post-harvest processing

After harvesting the fresh cherry is transported to a processing station, where it is weighed and recorded for payment and/or accountability, if this has not already been done. Regardless of the subsequent processing steps, the cherry then receives its first level of sorting. The cherry is spread onto the floor, which is often covered with a tarpaulin, or something smooth and relatively clean, and rapidly sorted by hand to remove the worst green (unripe) and very dark (over-ripe) fruits. Badly bruised and other defective fruits are also removed, along with any twigs (branch shoots), leaves, and other unwanted material (particularly stones, metal objects, insects and other small animals, etc.). Rejected cherry, and particularly under-ripe and over-ripe fruit, is often collected together and used in the households of the coffee workers. Once this first sorting has taken place, the coffee is ready for one of the two major steps in processing: either dry (natural) or wet (washed) processing.

Dry-processed coffees predominate in Ethiopia. Wet processing is common in the Jimma-Limu coffee area, western Sidamo (including Yirgacheffe), and suitable localities in the Kaffa, Illubabor and Wellega coffee areas (Figure 7, page 49), although isolated washing stations occur throughout Ethiopia; further information is provided in Chapter 10. The use of small-scale, low-cost, wet-processing systems, notably using the Penagos® depulper, allows wet processing in areas that have good suitability for high-quality coffee, but lack the resources

Sun-dried coffee fruits (dry cherry) Image: Alan Schaller (courtesy of Union Coffee)

to make significant investment in processing. The other major attribute of the Penagos® systems is that they use relatively small amounts of water, and can thus be placed some distance from rivers or major sources of fresh water.

Dry (natural) processing

Dry (or natural) processing represents the age-old, traditional methods of processing coffee in Ethiopia, and in many places outside Ethiopia. An estimated 75–80% of Ethiopia's coffee is processed in this way, requiring a low level of investment in terms of equipment and labour. Moreover, the requirement for low rainfall and low humidity during the drying of coffee cherry suits the climate of Ethiopia, which has a long, marked dry season that coincides with harvesting and processing. After harvesting and initial sorting, the fresh coffee fruits are spread out in a thin layer (2–6 cm deep; depending on the volume of coffee) outdoors, in the open air, to dry 'naturally' in the sun. The coffee is either dried directly on the (dry, earthen) ground, on woven mats, on tarpaulins or even on the tarmac roadside, or, preferably, on specially constructed concreted areas (patios) or raised drying beds (often called African beds). African beds consist of wooden frames covered with mesh of some description (made of natural materials, plastic, or wire), which holds the coffee about 1 m off the ground. The raised mesh considerably improves air circulation and thus drying, and keeps the coffee out of the reach of farm animals (chickens, goats and dogs) and sources of foreign matter. Irrespective of the where the coffee is dried, it is turned frequently to enable proper drying and to avoid mould and fermentation. When the cherry is dried on concrete patios, large wooden rakes or pushers are used to turn the coffee. Plastic sheeting may be used to protect the drying fruits from rain, and even night-time humidity and morning dew, especially when raised beds are used. After 2–4 weeks of drying, depending on the weather conditions and other factors, the fruits blacken and the pulp shrinks and hardens, and the drying process is complete. Throughout the drying process the moisture content of the drying cherry is monitored, preferably with a moisture meter. The cherry is dried so that it has a moisture content of between 12–20%. In Ethiopia it is quite common to see patio and ground-dried dried coffee covered with a white or greyish covering of mould. This influences the quality but if kept to a minimal level the mould should not excessively influence the taste of the coffee.

Wet (washed) processing

Wet processing of coffee is a relatively modern method, which probably first started in Ethiopia in the 1950s. Only in the past few decades has it become more widely established, mostly being undertaken in areas of good productivity and and where high-quality coffee is required. In this process, pre-sorted, freshly harvested cherry is first depulped, using depulping machinery (the wet-mill), whereby the fleshy, sticky part of the fruit (the pulp) is mechanically stripped from the coffee bean in combination with a flow of water. Once stripped of the pulp, the parchment coffee (i.e. with the parchment (endocarp) still present) remains covered with a thin layer of the pulp (often referred to as mucilage), and

Pulped coffee in the depulper Image: Jenny Williams

this has to be removed by fermentation. For fermentation, the parchment coffee is diverted from the depulper into concrete fermentation tanks into which water (usually river water) is added. Fermentation usually takes 12–36 hours, the exact time depending largely on the temperatures of the air and water, but also on the type of coffee being produced (see below). Fermentation breaks down the remaining sugar-rich mucilage attached to the parchment through bacterial (microbial) action, leaving the parchment free of pulp. After fermentation the beans are given two to four short washes using clean water, which removes the decomposed mucilage and any foreign matter. The parchment coffee is then placed onto small purpose-built drying racks for further hand-sorting, and afterwards spread onto raised drying beds (African beds; see above) for drying. Like dry-processed coffee, the drying coffee needs to be frequently turned and monitored to make sure that it is dried to the correct moisture level, and that the drying is neither too slow nor too fast. Generally, the parchment is dried more rapidly in the earlier stages of drying, but not so fast that the parchment cracks. Drying times vary depending on weather conditions and drying bed management, but the usual drying period is 10–15 days. On some farms, moisture meters are used to monitor the moisture content, which for the dried parchment should be 10–12%. The dried parchment varies from a pale straw to pale yellowish colour.

Dry vs. wet processing

Generally, wet-processed coffees are considered to be of a superior and more consistent quality compared to dry-processed coffees, although this matter is regularly debated by roasters, baristas and coffee-drinkers. Washed coffees are generally considered to be 'cleaner' and more delicate in the cup (i.e. during drinking), lighter-bodied, and with a brighter, better-defined acidity and brighter fruit notes. Washed coffees are also often said to be more balanced (i.e. lacking particularly strong tasting notes or biases). Dry-processed (natural) coffees tend to be fuller bodied, fruitier, sweeter

Coffee washing station

Image: Emily Garthwaite (courtesy of Union Coffee)

and less acidic, although in some cases the acidity is more pronounced and easier to define. Choosing between the two types often comes down to personal preference. Some drinkers prefer the more characterful and interesting (e.g. 'funky') natural coffees; and others prefer the cleaner, and usually more balanced characteristics of washed coffees. Dry-processing can offer taste experiences not found in washed coffees, such as distinct fruit and floral notes, although sometimes these characteristics are considered to make a coffee unbalanced and even faulty.

The extra steps involved in washing also mean that wet-processed coffees usually have fewer cupping defects. For example, unripe beans float in the fermentation tanks and can be removed easily, although this step can also be applied to dry-processed coffees. Generally, if dry-processed coffees are processed to a high standard, then they can match washed coffees in terms of quality and consistency.

One consideration for the consumer is that dry-processed coffees are more environmentally friendly, as the processing requires far less equipment, less fuel (for the depulping machinery) and most importantly little or no water. These factors make dry-processed coffee more carbon efficient, an important consideration as global coffee production produces a significant carbon footprint. In general, agriculture is a significant contributor to greenhouse gasses. Moreover, there can be some serious pollution issues resulting from wet mill run-off into streams and rivers, although properly implemented settling tanks (where waste water is stored and dispersed, rather than being released directly into the rivers) are helping to reduce this negative impact. Dry-processed coffee represents around 80% of total Ethiopian coffee production, although washed coffee is definitely on the increase, particularly following large-scale investment in the main coffee-growing areas.

Other processes (pulped natural and honey process)

Pulped natural and honey processing represent relatively new developments in coffee processing, and are undertaken to produce high-quality, speciality coffees. Pulped natural and honey processing are very similar. After the depulping stage, as for washed coffee (see above), the sticky parchment coffee is either given a quick rinse or not, but importantly the fermentation stage is omitted. Thus, when the depulped coffee is laid out on the drying beds it is still covered in the plentiful remains of the pulp (mucilage), the amount depending on the level of depulping and subsequent washing. Once on the drying beds, it is dried in a similar way to washed coffee. The honey process differs from pulped natural by the increased amount of pulp that is left on the parchment, which is controlled mainly by the amount of water that is used. On the drying bed, both types are instantly recognisable by the light to golden brown dried, sweet, sticky coating on the parchment (more so in the honey process). The coating (dried mucilage) contains high levels of sugars. The coffee is fully dried with the coating in place, again similar to fully washed coffee. The dried mucilage is only removed at the hulling stage, along with the parchment. Processing the coffee in this way adds sweetness and specific flavour characteristics, and if done correctly produces sweet-tasting, interesting coffees, with qualities that usually set them apart from natural or washed coffees. Varying the amount of mucilage left on the parchment influences the characteristics of the coffee. Pulped natural and honey-process coffee is not common in Ethiopia, but production is underway on specific farms and cooperatives.

There are still a large number of other factors that influence quality and taste, even after the green coffee has been through most of the processing steps. These factors include pre-shipping processes (e.g. cleaning, sorting and grading), storage, and transportation; and, of course, in the country of consumption, the way the coffee is roasted, made and served, also has a considerable influence.

Resting

After drying, both natural (dry) and washed (wet) processed coffees require a 'resting' period. For dry processed coffees this is undertaken after the cherry has been properly dried, and for wet processed coffee at the dried parchment stage. The resting step lasts for 1–3 months, preferably in a properly ventilated warehouse or other suitable storage facility. In the case of small-scale farmers, the coffee is often rested in the smallholder's house or outbuilding. Resting or 'maturation' is undertaken to improve cup and transportation quality, but the exact duration can be more influenced by transport and processing logistics rather than careful timing. Sometimes the coffee is stored for longer periods, such as more than 6 months, but then the coffee starts to 'age' and will develop different flavour characteristics. If the moisture levels of the stored coffee are too high, or if there are other negative influences (heat, proximity to strong odours, or extended storage), the coffee will develop negative qualities (faults). In some countries coffee is aged intentionally to develop specific flavour characteristics, but this is not the case in Ethiopia.

Hulling

After drying, both dry- and wet-processed coffees need to be hulled. For dry-processed coffees the hard, blackened part (the husk, which was once the fleshy pulp of the fruit) has to be removed. For wet-processed coffee the parchment is removed by hulling (sometimes referred to as milling). Hulling is a mechanical process, performed using a machine of various manufacture, and usually driven by a petrol or electric motor. The facility that undertakes hulling is usually called the milling station. The hulling machines rub off the husk, and/or parchment, leaving clean, green coffee. After hulling the coffee may undergo sorting to remove defective beans and foreign material, especially for dry-processed coffees, and this is usually done by hand. Traditionally, hulling of coffee for household use was undertaken using a large, wooden pestle and mortar, and this method still persists in Ethiopia. Industrial hulling processes produce a lot of waste, especially for dry-processed coffees. The waste can be used for agricultural purposes, such as mulch and soil conditioner. In most Ethiopian milling stations, however, hulling by-products produced from dry-processed coffee are left to form large heaps next to the milling station, which eventually decompose and, due to the heat produced during decomposition, smoulder away for many weeks or months.

Green coffee

Green coffee represents the state at which coffee is exported to the consuming countries, for roasting, and then packaging, distribution and consumption. A large proportion of green coffee is not exported, however, and is roasted in Ethiopia, by the householder for home consumption and commercially for the domestic market.

Green coffee can be various shades of green, including bluish green to blue-green, which is often considered the best quality, through to greyish blue, green or greenish, to almost white/

Coffee hulling station

Image: Jenny Williams

Green coffee, after hulling Image: Jenny Williams

very light brown (lowest quality)[49]. Generally, washed coffees are brighter and lighter in colour than dry-processed (natural) coffees. In the pre-export grading system (see below) the beans may also be referred to as faded, or blackened, and both represent faults in the coffee. Faded beans are soft and have a light brown, whitish brown, grey, or pale brown-green colour. The causes of faded beans are various, including over- or under-drying, poor storage (e.g. high humidity, high temperatures), or over-storage (kept too long in the storage facility). Blackened beans are brown to black in colour, and may be caused by drought, late-harvesting (over-mature), or over-fermentation during wet-processing. Other green coffee defects include 'cuts' (caused by suboptimal adjustment of the depulper, for washed coffees), insect damage (holes in beans), and a shrunken appearance or an irregular shape (often a sign of drought).

When green coffee reaches the delivery centres or warehouses it may undergo further sorting and cleaning, either using machinery (a colour sorter or airflow machine) or in many cases by hand and eye, to remove any hulling remains, foreign particles, and defective beans. In some cases further milling may take place to further clean the coffee. The green beans are also size-graded using a series of grids or sieves.

Speciality coffee

Speciality coffee is a classification given to high-quality coffee, which must be able to pass very specific physical grading criteria and sensory (aroma and taste) evaluation tests (cupping). The Speciality Coffee Association (SCA) defines speciality coffee in its green (unroasted) stage as: '...coffee that is free of primary defects, has no 'quakers' [an unripe and often bad-tasting coffee bean], is properly sized and dried, presents in the cup free of faults and taints and has distinctive attributes'. A speciality coffee should also be naturally sweet, and possess interesting flavour characteristics. To evaluate coffee, the SCA uses a standard international Quality Score. Only coffees with 80 or more points (out of 100) can be classified as speciality coffee. Coffees with a score close to, or above 90, are exceptional.

Grading

Before export all coffee must be graded. The Ethiopian Commodity Exchange (ECX) undertakes grading prior to accepting coffees for storage in its warehouses. Grading may also take place outside the ECX. Coffee grading systems are strict and based solely on quality. Under the ECX grading system there are five main grades (1–5) and an under-grade (UG). Grade 5 is divided into three subgrades (5A, 5B, 5C). The first two grades are prefixed with Q: Q1 has a cup quality score of equal to, or greater than, 85, and for Q2 the score is 75–84. The grading system uses the combined result of: (1) physical features (raw value) of the green coffee (40% of total value), and (2) cup quality or liquor characteristics (cup value) (60% of total value). For washed coffee the raw value (40%) is made up of defects (20%), shape and make (5%), colour (5%) and odour (10%); and for unwashed coffee, defects (30%) and odour (10%). For washed and unwashed coffees, cup value (60%) is made up of cup cleanness (15%), acidity (15%), body (15%) and flavour (15%). In addition, the quality evaluation includes an assessment of green bean moisture levels. Typically, beans with greater than 12% or below 9% moisture levels are rejected. Further information on the ECX grading system can be found within the ECX Coffee Contracts document[50].

Under the conventional grading system used in Ethiopia, nine grades are used, covering both unwashed and washed coffees, from Grade 1 (highest quality) to Grade 9 (lower quality), and Ungraded (lowest). Grades 1 and 2 represent speciality coffee grade (as above); Grades 3–9 are export-quality coffees, and Ungraded coffee generally stays in Ethiopia for the home market. Even though unwashed coffees are sometimes considered to be of lower quality, they can achieve Grades 1 and 2, as speciality coffees.

The trading system and coffee classification

There are various ways that processed (and unprocessed) coffee is traded within Ethiopia before reaching the central trading system (see below). Processed coffee (and fresh cherry) can be taken to the local market or sold to cooperatives or coffee middlemen (*akrabis*). Generally, coffee exporters buy green coffee from the larger-scale regional suppliers, including the cooperatives, through the Ethiopian Commodity Exchange (ECX) auction system (as part of the regulatory requirement for in-country coffee transactions), although limited exportation from cooperatives and other producers may go through other routes. The ECX auction system is strictly organised in order to ensure that in-country value-chain steps are controlled, and are measurable and traceable. For example, there is a regulatory requirement forbidding trade or movement of coffee from one coffee-growing region to another. All regional coffee producers and suppliers are required to deliver their coffee directly to the dedicated ECX station within their coffee-growing area[50], where the coffee is stored until it is transported to ECX's national coffee exchange market in Addis Ababa. This allows traceability of origin, at least to one of the major coffee-growing areas, or a specific larger-scale origin (e.g. Yirgacheffe, Guji). This system is enforced by check-points on major roads, and by origin and quality inspection at various stages of the transportation and transaction process, including checks by government officials. One of the key activities for quality control is grading, which is undertaken by the ECX at the federal and national level (see above).

In order to organise and manage the traceability of the various origins, the Ethiopian Commodity Exchange (ECX) has devised its own coffee classification system, based on geographical growing areas and their cup profiles, factoring-in the main processing types (i.e. natural/dry vs. washed/wet-processed) and grading. This classification system[50] provides a framework for a number of logistic processes, and allows traceability should the exporter or buyer want to know the area from where the coffee was grown (origin). As an origin is often linked to a specific cup profile, the coffee buyer or consumer will know what to expect from any particular coffee, and may select an origin to suit one or a number of purposes. For many, however, and particularly for speciality coffee buyers, this system does not provide sufficient traceability: coffee from numerous farms, and from across more than one origin, can be combined into a single lot, for example. The ECX system has undergone significant change over the past few years, and further changes are likely in the near future.

Exportation

After all processing and post processing steps have been completed, the coffee is weighed and packed into 60 kg food-grade jute sacks, or is bulk-containerised into a lined shipping container. A single 20 ft shipping container holds 18,000 kg (18 metric tonnes (MT)) of coffee in 300 × 60 kg sacks; or when loose and un-bagged, 19,000 kgs (19 MT). From Addis Ababa, coffee loads are transported by road or rail to the Port of Djibouti, where they are delivered to the designated shipping carrier for export via either the Gulf of Aden or the Red Sea. The time taken from harvest to delivery at the buyer's warehouse in the destination country can be lengthy. Coffee harvested in Ethiopia in November might take 9–10 months to reach the docks in the United Kingdom, and thus not reaching its destination until July or August the following year, for example.

Sorting green coffee by hand (Yayu, Illubabor coffee area, south-west Ethiopia)

Image: Jenny Williams

Wild coffee forest near Bonga
Image: Jenny Williams

10. The coffee areas

Overview

The coffee landscape of Ethiopia is organised here into five zones, and 16 areas (Table 1, Figure 7). The main coffee-growing areas are found within the Oromia Region and the Southern Nations, Nationalities, and Peoples' Region (SNNPR), with modest production in the Amhara Region and minor output in the Benishangul Gumuz Region (Figure 7). The five coffee zones are: North, South West, Rift, South East and Harar. The greater part of coffee production in Ethiopia is from the largely forested South West coffee zone, followed by the Sidamo (including Yirgacheffe) coffee area of the South East coffee zone. Harar was once an important producing zone but has declined considerably since the 1960s, and possibly some decades before. The other coffee zones (Rift and North) are modest producers compared to the South West coffee zone and Sidamo coffee area.

The 16 coffee areas (Table 1) are based on modelled coffee (niche) suitability and forest cover (from satellite data; see Chapter 2), incorporating common conventions and terminology from the coffee sector, and terminology adapted from the Ethiopian Commodity Exchange (ECX) contract system[50]. It was not possible to follow the ECX system precisely, as the designation of the ECX geographical nomenclature also depends on the method of processing (e.g. washed vs. unwashed; speciality vs. commodity), and is thus not strictly aligned with geographical distribution. Moreover, the ECX contracts may change from year-to-year. Table 1 provides a summary of the 16 coffee areas, indicating the general position within Ethiopia (relative to the Rift Valley), their assigned coffee zone, and the administrative region in which each coffee area is situated. Figure 7 shows the geographical position of the coffee zones and coffee areas. Figure 6 (page 29) summarises the present-day suitability for coffee-growing and recent forest change. The Atlas map pages (1–41) show the suitability for coffee farming, in relation to forest and non-forest cover, in detail.

In the section below we provide a short narrative for each of the coffee zones and their coffee areas, giving basic location details and an overview of the main production systems and their characteristics.

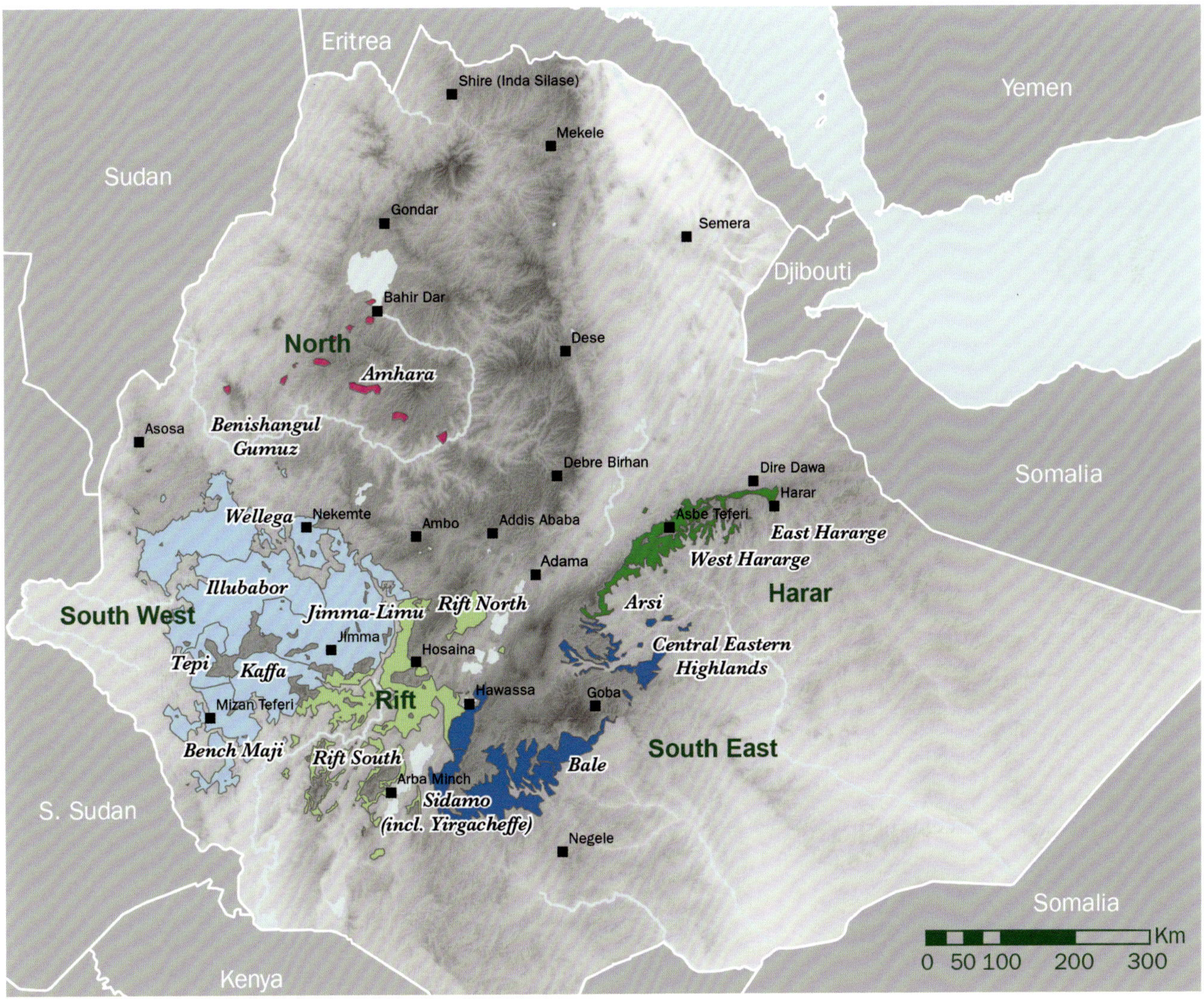

Figure 7. The coffee-growing zones (North, South West, Rift, South East, Harar) and coffee areas (in italics) of Ethiopia. See Table 1, on page 50.

Position in Ethiopia	Main coffee zone	Coffee area	Region (administrative)
A. West of Rift Valley	A1. North	A1.1 Amhara	Amhara
		A1.2 Benishangul Gumuz	Benishangul Gumuz
	A2. South West	A2.1 Wellega	Oromia
		A2.2 Illubabor	Oromia
		A2.3 Jimma-Limu	Oromia
		A2.4 Tepi	SNNPR and Gambella*
		A2.5 Kaffa	SNNPR
		A2.6 Bench Maji	SNNPR
B. Rift Valley	B1. Rift	B1.1 Rift North	SNNPR and Oromia*
		B1.2 Rift South	SNNPR
C. East of Rift Valley	C1. South East	C1.1 Sidamo (incl. Yirgacheffe)	SNNPR and Oromia*
		C1.2 Bale	Oromia
		C1.3 Central Eastern Highlands	Oromia
	C2. Harar	C2.1 Arsi	Oromia
		C2.2 West Hararge	Oromia
		C2.3 East Hararge	Oromia

Table 1. Main coffee zones and coffee areas and their position within Ethiopia. See Figure 7, page 49.

Notes: *Only a minor part of this region is within this coffee area. SNNPR = Southern Nations, Nationalities, and Peoples' Region.

A. West of the Rift Valley

A1. North Zone

This is a minor coffee zone, with isolated areas of coffee production. It is divided into two coffee areas: Amhara and Benishangul Gumuz, each of which sits within their respective administrative regional boundaries (Table 1). Generally, the climate in the Amhara and Benishangul Gumuz regions is too dry for coffee (the wet season is shorter than that in the south-west), but where there is sufficient water for irrigation and/or land available at higher elevations, coffee is grown successfully. Most of the coffee in the North coffee zone is shade coffee. Arabica coffee is not wild (indigenous) in this zone.

A1.1 Amhara (Atlas map pages 26, 27, 33, 34)

The Amhara coffee area starts around the town of Debre Markos, where coffee gardens are the predominant production system. Farms around the town of Finote Selam contribute significantly to the coffee production of the Amhara coffee area. The high water table and presence of rivers and streams in and around Finote Selam enables irrigation, and there are some sizeable farms dedicated to irrigated production. Further north there is small- to medium-scale coffee production between the towns of Injibara and Chagni, with coffee being grown both with and without irrigation. Throughout this area, where conditions are suitable, there are numerous small coffee gardens. Zege (Zeghie) Peninsula (1,800–1,950 m), on the south-western edge of Lake Tana, is a small patch of humid forest that has been producing coffee for at least the past two centuries. Many hundreds of small-holders produce semi-forest coffee (SFC) at Zege, but good harvests are infrequent. Coffee is not indigenous to Zege, but is believed to have been brought from south-west Ethiopia 200–300 years ago. The locals consider their coffee to be milder than that coming from south-west Ethiopia, with some referring to it as *Zege buna*.

Zege is often considered the most northern limit for coffee in Ethiopia, although there are isolated coffee gardens north of Bahir Dar. Historical records from the eighteenth and early nineteenth centuries report limited coffee cultivation (probably coffee gardens) from Gonder (Amhara Region) and the Tigray Region. Coffee-growing in the Amhara Region is generally restricted to 1,700–2,200 m, although new coffee production areas are being established at elevations in excess of 2,400 m. Production is largely restricted to natural (dry) processing, with isolated washing stations. Amhara coffees usually have light acidity and light to medium body, with notes of red berries and citrus; Zege coffee can have interesting fermented fruit characteristics, with notes of port, sherry and whisky.

A1.2 Benishangul Gumuz (Atlas map page 25)

Benishangul Gumuz is the smallest and most isolated of all the coffee areas, and is restricted to a small high plateau at Debre Zeyit (Wembera). Coffee farming occurs on the western side of the plateau, west of Debre Zeyit town, at 1,700–2,200 m. Coffee-growing at this location is mostly above 1,800 m, on steep slopes, and of the forest garden coffee (FGC) type, with irrigation used where water is available. Coffee from this area is not well known, and mostly not exported. The coffee is dry-processed, and covers low to high grades. Coffee from Debre Zeyit has some excellent qualities (silky body, notes of citrus, various fruits and tropical flowers, and good sweetness). The positive attributes are assisted by high elevation, irrigation and good-quality harvesting. Coffee is not indigenous to this area and the plants used in this area probably originate from Wellega or Jimma-Limu coffee areas.

A2. South West Zone

The South West is the main coffee-producing zone of Ethiopia, located on the western side of the Rift Valley, on the massive highland plateau in the south-western part of the country (Figure 8, page 61). The South West coffee zone includes the famous coffee localities of Bonga, Kaffa and Limu, as well as the well-known coffee town and delivery centre of Jimma. The forests of the south-west are home to the largest populations of wild (indigenous) Arabica coffee. The South West zone is divided into six coffee areas: Wellega, Illubabor, Jimma-Limu, Tepi, Kaffa, and Bench Maji.

A2.1 Wellega (Atlas map pages 16–18, 25)

The Wellega coffee area can be subdivided into three smaller areas, from west to east: Kelem Wellega, West Wellega and East Wellega. Kelem Wellega is the western-most part of the area, and includes the towns of Dembi Dollo, Ashi (Sedi), Gidami and Mugi. Garigeda lies between Ashi and Gidami, and includes both forest coffee production areas and natural coffee forests (with wild Arabica coffee). The woreda name Anfilo (or Anfillo) is sometimes associated with coffee from this area. Kelem Wellega includes many very good coffee-growing localities, particularly those with good forest cover at 1,700–2,100 m; forest in this area continues upwards to much higher elevations (over 2,300 m). Kelem Wellega is not a well-known coffee origin but has the potential for high-quality (speciality) coffee, with some distinct and unusual flavour profiles. West Wellega includes the major town of Gimbi, and the smaller towns of Guliso and Alem Teferi, with much of the coffee-growing around these three towns at 1,300–1,700 m. The area comprises coffee gardens and small to large coffee farms, as well as some smaller-sized plantations. The coffee is grown in full sun, light shade or shade, and there are also small, isolated patches of forest coffee (mostly forest garden coffee (FGC)). East Wellega includes the large town of Nekemte, and is similar in its coffee cultivation characteristics to West Wellega but with less coffee production (the eastern part of Wellega is generally less suitable for coffee-growing).

The Wellega area produces washed and dry-processed coffees, covering all grades[50]. Coffee from the Wellega area is often referred to as Lekemti, Nekemti (after the town of Nekemte) or Gimbi, but it could originate from almost anywhere in the Wellega coffee area. The higher-grade coffees can be distinct, sometimes with honey, citrus, and floral notes, especially those from Kelem Wellega. Coffee from East and West Wellega, particularly those referred to as Lekemti, tends to have rather large, elongated beans.

A2.2 Illubabor (Atlas map pages 17, 18)

The Illubabor coffee area is positioned south of the Wellega coffee area and north of the Tepi, Kaffa and Jimma-Limu coffee areas. The major town of the area is Metu (or Mettu), with the large towns of Gore and Bedele to the west and east, respectively. The Illubabor area extends some considerable distance to the west, to the west of Gore towards Gambela, until the high plateau ends. The southern part continues south of Bedele, and includes coffee farms around the town of Gechi. Large parts of this area are forested, including the dedicated coffee forest reserve at Yayu. Yayu and other forests in the area are home to important populations of wild (indigenous) Arabica coffee, which (as in many other areas of Ethiopia) are used to provide planting stock for coffee production. The coffee produced in this area is mostly of the forest (FC), semi-forest (SFC) and forest garden coffee (FGC) types, although there are some large and intensively planted and managed areas, including small plantations. Often overlooked by coffee buyers and consumers, this area has much to offer and is now producing some excellent washed and natural processed coffees, particularly at elevations at and above 1,500 m. The coffees are generally low in acidity and body and display notes of tropical fruit, stone fruit, chocolate, caramel, and vanilla. Within the ECX coffee system, Illubabor coffee mostly falls under the Limu (washed) and Jimma (unwashed) contracts.

Inspecting the coffee harvest (Debre Zeyit (Wembera), Benishangul Gumuz coffee area, north-western Ethiopia) Image: Aaron Davis

A2.3 Jimma-Limu (Atlas map pages 10, 18, 19)

The Jimma-Limu coffee area is in the eastern part of the South West coffee zone, and is centred on the town (and coffee delivery centre) of Jimma. It can be subdivided into two parts: Jimma and Limu. The Jimma part includes areas north of Jimma town, through the towns of Yebu, Gembe, Agaro, Toba, and to Dembi and Yembero. To the south of Jimma, the area extends to Dedo. To the east of Jimma it reaches Welkite, although this part of the area is largely dominated by scattered coffee gardens. The south-western part of the area includes the towns of Seka and Sembo, ending before the town of Shebe. The north-western part of the area extends up to the edge of the Illubabor coffee area, while the south-western part adjoins the Kaffa coffee area. There are important coffee-growing locations west and south-west of Agaro, and to the south around the towns of Beshash and Chira. The elevation range is 1,500–2,200 m and sometimes almost up to 2,300 m. The Limu part of the Jimma-Limu coffee area is located north-east of Jimma and east of Agaro, and includes the towns of Ambuye, Gent (Suntu) and Seka (en route to Atnago), at elevations of 1,700–2,100 m.

Jimma-Limu is a large and important coffee-producing area. Most of the production is forest garden coffee (FGC), semi-forest coffee (SFC), and there is also the large, long-established Limmu Coffee Farm plantation. Despite the close association between Limu and Jimma, the coffees from these two localities can vary considerably. Limu carries a reputation for producing the better coffees, possibly because of the larger number of washing stations and generally better climate, although Agaro (Jimma) also carries a deservedly high reputation. Washed coffee from Illubabor (see above), and further afield, is sometimes referred to and/or classified as Limu. This sometimes causes confusion, although the flavour profiles can be similar. Limu is noted for its washed coffees, and for dry-processed coffees of higher grades. The standard of agronomy and processing is good in many places, with some well-established larger farms. Limu coffee is renowned for its balance, sweetness, lightly spiced, pleasant wine-like characteristics, and light to medium body; some Limu coffees have distinct floral and stone fruit notes. Coffee from Agaro, which is in the Jimma part of the Jimma-Limu coffee area, also has good sweetness and subtleness, with notes of stone fruit (apricot, peach), citrus, and sweet-scented tropical flowers.

A2.4 Tepi (Atlas map pages 9, 17)

The Tepi (or Teppi) coffee area is positioned between the coffee areas of Illubabor and Bench Maji (Figure 7, page 49). It starts just north of Tepi, around Gecha, and reaches beyond Masha towards Gore, and to areas west and east (in the southern part of the area) of these towns. There is a considerable amount of high ground in the Tepi coffee area, especially between Gecha and Masha (2,000–2,600 m), although much of this ground is too high for coffee (above 2,200 m). In Tepi, coffee is mainly found at 1,700–2,100 m. Much of the area is still densely forested, with some important populations of wild Arabica coffee. The Tepi Coffee Plantation is located to the west and north-west of Tepi, and is one of the largest coffee plantations in Ethiopia, covering over 8,000 ha. The Tepi plantation covers elevations of 1,200–1,900 m, although the majority of the coffee is situated below 1,500 m. The coffee from this area is largely of the forest (FC), semi-forest (SFC) and plantation coffee (PC) production types. Tepi coffees often have low acidity, slight to medium body and a soft, balanced flavour.

A2.5 Kaffa (Atlas map pages 9, 10)

This famous coffee area, often said to be the 'home of coffee', is centred on the town of Bonga. To the west, this coffee area extends through the towns of Wushwush (the site of large tea plantations), Dara, and Dimbira, and ends around Wacha. To the east, the Kaffa coffee area reaches the town of Shebe. It also covers a large area north and south of Bonga, adjoining the Jimma-Limu coffee area to the north-east, and the Bench Maji and Tepi coffee areas to the west. The southern part includes the town and coffee-producing locality of Anderacha. In many places, there are still substantial tracts of natural

Harvesting coffee (Yayu, Illubabor coffee area, south-west Ethiopia)

Image: Alan Schaller (courtesy of Union Coffee)

Young coffee plants dwarfed by forest trees (near Gore, Illubabor coffee area, south-west Ethiopia) Image: Aaron Davis

and semi-natural forest, much of which is used for semi-forest coffee (SFC) production. The Kaffa coffee area includes the large and globally important UNESCO Kaffa Biosphere Reserve, which aims to preserve the wild coffee forests and their wildlife, as well as the considerable natural genetic resources of wild and semi-wild Arabica coffee found there. The Kaffa area produces washed coffees, particularly around Bonga and Anderacha, as well as dry-processed coffees. The major coffee production systems are of the semi-forest (SFC) and forest garden coffee (FGC) type, as well as small plantations, over a wide elevation range of 1,300–2,200 m, with most coffee at 1,700–2,000 m. Kaffa coffee has a medium body and acidity; the flavour profiles and specific notes are rather poorly known and require further assessment.

A2.6 Bench Maji (Atlas map pages 2, 9)

This area includes the town of Mizan Teferi (Mizan Aman). The south-eastern part of the Bench Maji coffee area encompasses Bebeka and Gura Ferda, and in the southern part it includes the famous coffee origin of Gesha (east of Jemu) and areas around the town of Maji, reaching the northern edge of the Omo National Park. Bench Maji is an area of considerable contrasts, from the highly managed coffee plantations of Bebeka (1,000–1,400 m), to the largely untouched forests of Gura Ferda (1,700–2,300 m). It also includes the remnant wild coffee forests of Gesha (1,800–2,000 m) and the isolated coffee forests of Maji (mostly above 2,000 m). In the southern part of the Bench Maji area, coffee cultivation is interspersed with lower-lying crop and pasture lands. There are some important populations of wild Arabica in the forested areas around Gura Ferda, Gesha and Maji, and also at Sheko (south-west of Tepi), a densely forested area that also produces forest (FC) and semi-forest coffee (SFC).

The main production systems of the Bench Maji area are forest garden coffee (FGC), semi-forest coffee (SFC), and plantation coffee (PC). Bebeka is currently the largest coffee plantation in Ethiopia, covering around 10,000 ha, producing more than 2,000 metric tonnes of coffee, and employing about 2,000 permanent workers. It produces Arabica coffee, some robusta coffee (*Coffea canephora*), and also fruit, spices and honey.

Washed coffees of speciality, and high to lower grades, are produced in Bebeka and other parts of Bench Maji. Dry-processed coffees are produced throughout the area, from speciality through to low grades (and under-grades)[50]. Gesha is the home of the world-famous Gesha (Geisha) coffee, and this origin is now starting to produce its own very high quality washed, semi-washed, and dry-processed coffees. Gesha (Geisha) coffee grown within and outside Ethiopia (e.g. in Panama) is noted for its outstanding aroma and flavour, with notes of jasmine, black tea, and tropical fruit, and for profound sweetness. For these reasons, Gesha (Geisha) coffee is sought-after and usually commands a high price. There are other distinct flavour profiles across the Bench Maji coffee area and these require further assessment and characterisation.

B. Rift Valley

B1. Rift Zone

The Rift coffee zone is located between the main south-western plateau and the Sidamo coffee area (Figure 7, page 49). Coffee is grown where conditions are wet and cool enough, at suitably high elevations above the low lying (and considerably warmer and drier) lands of the Rift Valley. The coffee landscape here is dominated by garden coffee (GC) and coffee cultivation in small, mixed-cropping systems. The Rift Valley is not a major producing area, and is largely devoid of forest. It is divided into two parts: Rift North and Rift South. Coffee from the Rift falls within the Sidamo C and Sidamo E contract system of the ECX, with Soddo as the coffee delivery centre. The Rift Valley coffee zone produces mainly, but not exclusively, dry-processed coffee. Bean size and shape is variable but tends to be small, rather rounded and hard (dense), especially for beans grown in the Rift South coffee area, and is thus similar physically to Yemen Mocha (Mocca) coffee. The physical characteristics of the beans are probably due to climatic factors, particularly lower rainfall and higher temperatures.

B1.1 Rift North (Atlas map pages 11, 19, 20)

Soddo is the main town of the Rift North coffee area. Coffee-growing here is fragmented and spread over a large area, with moderate to high concentrations of garden coffee (GC) around the towns of Durame and Angecha (north of Soddo) and Tulema and Waka (west of Soddo), in locations south of Welkite, and in the northern-most part of the area around Butajira. The coffee-growing areas are largely confined to elevations of 1,800–2,200 m.

B1.2 Rift South (Atlas map pages 3, 4, 11)

The Rift South coffee area is located in the southern part of Ethiopia's coffee landscape, south and west of Lake Abaya. Arba Minch is the principal town in this area. As in the Rift North area, coffee is confined to higher elevations (1,700–2,200 m), for example to the west (Jinka, Wib Hamer, and Bulki (Bulchi)) and south (around Gidole) of Arba Minch.

C. East of the Rift Valley

C1. South East Zone

The South East is a major coffee zone, located on the eastern side of the Rift Valley. The zone incudes the well-known Sidamo coffee area, and the famous coffee town of Yirgacheffe. The zone is divided into three coffee areas: Sidamo, Bale, and the Central Eastern Highlands. Most of the production in the South East zone is confined to Sidamo, particularly the western part, with far less coffee farming in Bale, and only minor production in the Central Eastern Highlands. The eastern part of the Sidamo coffee area includes many of the lesser-known coffee origins, such as Bensa, Bona Zuria, and West Arsi (Nansebo) and Guji, although Guji has recently, and quickly, earned a reputation for unique, high-quality coffee.

C1.1 Sidamo (including Yirgacheffe) (Atlas pages 4, 5, 11, 12)

The well-known Sidamo (or Sidama) coffee area of south-east Ethiopia includes the equally renowned coffee locality Yirgacheffe (Yirga Chefe). Sidamo is a major coffee-producing area, and encompasses an incredibly diverse range of flavour

Semi-forest coffee production (Harenna Forest, Bale coffee area, south-east Ethiopia) Image: Aaron Davis

profiles. The main part of the Sidamo coffee area is narrow and discrete, running along a broad rift, north to south, from Hawassa (Awassa) to Gedeb (just south of Yirgacheffe). The northern part of this main area includes the towns (from north to south) of Wendo Genet, Yirga Alem and Aleta Wendo. In this part of Sidamo, Hawassa is the main coffee delivery centre. In the southern part of the main Sidamo area, Dilla is the main town and coffee delivery centre. Towns include (from north to south) Wenago, Koti, Yirgacheffe and Fiseha Genet (Abel). Under the ECX contract system, coffees from this western part of the area mostly fall under Sidamo B and Yirgacheffe A/B, although there are also other ECX classifications linked to origin (e.g. Wenago, Kochere, Wonsho). Small villages around Yirgacheffe are also sometimes listed as specific coffee-producing origins.

Beyond the main western part of the Sidamo coffee area, eastern Sidamo extends into the western part of the Bale Mountains. The road from Aleta Wendo, through Bore, Kibre Mengist, Zembaba Wiha, to Wadera, cuts through the south-eastern part of Sidamo. Further north (Atlas map pages 5 and 12), there are three lesser-known coffee localities: Bona Zuria, Bensa, and West Arsi (Nansebo). West Arsi (Nansebo) produces high-quality washed and natural coffees, at high elevations (1,800–2,300 m), based close to the town of Bulega. Under the ECX classification, coffees from this part of Sidamo usually fall under Sidamo A. In the most southern part of the Sidamo coffee area is the coffee locality of Guji, which extends over a considerable area and includes higher-elevation coffee-growing areas. The principal towns of this area are Agere Mariyam and Kercha (see Atlas map pages 4 and 5), with most of the production occurring in isolated localities at higher elevations of 1,900–2,300 m, including some larger farms.

Coffee farming in the western part of the Sidamo coffee area ranges in elevation from 1,600– 2,100 m, and in the eastern part 1,700–2,200 m, and sometimes as high as 2,300 m. The western part of the Sidamo coffee area is set apart from the rest of the Ethiopian coffee landscape by its distinctive cultivation system. Families in this area farm small and intensively managed mixed crops in association with forest trees or other canopy trees or vegetation, in a system known as agroforestry. The individual plots are typically a quarter to half a hectare and feature coffee, the starch staple (and banana-like) enset (*Ensete ventricosum*), yams (various species of *Dioscorea*) and eucalyptus (various species of *Eucalyptus*) among the main crop species. Some parts of Sidamo look heavily forested, until closer inspection reveals an intensive and productive agroforestry system. In other parts of the Sidamo coffee area, particularly north of Yirga Alem and Aposto, the cultivation system is more open, and either lightly shaded or in full sun.

The Sidamo coffee area produces washed and dry-processed coffees, with recent years seeing a large increase in the number of washing stations. Sidamo and Yirgacheffe coffees are noted for their extremely distinctive, often very fruity, floral and citrus flavour profiles. Most Yirgacheffe coffees have a particularly unique and distinctive flavour profile, with intense fruit, nuts, citrus, chocolate and almost perfume-like qualities, which is often referred to as the 'Yirgacheffe flavour'. Guji coffee is generally not well known but is quickly gaining a reputation for delicious, high-quality coffees (washed and natural), with notes of sweet citrus, floral, tea, rose, and various fruits, but especially vibrant, juicy blueberry. Even less well known are the coffees from Bona Zuria, Bensa and West Arsi (Nansebo). West Arsi (Nansebo) is now more evident in the speciality coffee market; the washed versions possess good balance, bright acidity and smooth mouthfeel, with notes of black tea, bright citrus, and floral aromas (e.g. cherry blossom).

C1.2 Bale (Atlas map pages 5, 6, 12, 13)

The Bale coffee area is situated in the southern part of the Bale Mountains, which have peaks of just over 4,300 m and a plateau area (Sanetti Plateau) with an average elevation of 4,000 m. Coffee is not farmed at these high elevations but within the Harenna Forest and adjoining forest areas. Here we find humid, evergreen forest of the same general type as that found in south-western Ethiopia (i.e. Moist evergreen Afromontane forest (MAF)[5]. Harenna forest is home to wild Arabica coffee, and it is these plants that are used to produce coffee, mostly of the forest (FC) and semi-forest (SFC) production types, but also forest garden coffee (FGC). Around 9,000 households are engaged in coffee production in Harenna Forest. Even though the forest extends up to 2,300 m, coffee production is confined to a belt from 1,400 m, where the humid forest starts, to around 1,800 m, the upper limit being restricted by low temperatures. The town of Mena (Dolo Mena) lies to the south of the main forest area.

The Bale coffee area is poorly known as a coffee origin, possibly because access was difficult until quite recently. Coffee from Bale is physically similar to some of the Harar coffees, with a large, long-oblong bean. Bale coffee has a distinctive flavour profile, including notes of floral, fruit, vanilla and sweet nuts. Bale coffee can be outstanding, although more investment is required to make the best of this potentially exceptional coffee. All of the coffee is currently dry-processed, and classified as Bale under the ECX contract system.

C1.3 Central Eastern Highlands (Atlas map pages 12, 13, 14)

This coffee area is situated south of the Arsi coffee area (Harar zone) and north of the Bale coffee area. The main towns of this area include Seru, Gaser, Ginir, Jara and Sheki Husen. The Central Eastern Highlands is a minor coffee area, largely comprising garden coffee (GC), grown under slight shade or in full sun. Coffee production occurs over a large area here, but production is slight compared to that of the Sidamo coffee area, with the majority of production for home use and the local market. Coffee from this area is dry-processed.

C2. Harar Zone

This well-known coffee origin sits on the north-eastern side of the Rift Valley, and runs along a mountainous ridge stretching from Chole (south-east of Adama (Nazret)) to the legendary city of Harar. The coffee-growing climate here is drier, and sometimes cooler, than that of the rest of Ethiopia's coffee landscape and, like the coffee-growing areas on the eastern side of the Rift Valley, has a bimodal rainfall pattern (see

Coffee-growing in the Harar coffee zone (Arsi coffee area, eastern-central Ethiopia) Image: Aaron Davis

Chapter 6). The natural vegetation is different to those of most of the other coffee areas, and is characterised by a drier type of vegetation, dominated by evergreen trees or shrubs, and common native coniferous species (classified as Dry evergreen Afromontane forest (DAF))[5]. Much of the coffee is grown without shade (sun coffee), or with minimal shade, mostly in coffee gardens (garden coffee (GC)), as part of a mixed cropping system. Some of the coffee in Harar is grown in terraces, and where there is access to water it may be irrigated.

The coffee from Harar is mostly dry-processed, although there are isolated washing stations. Harar coffees often have a distinct jam-like flavour, with a characteristic blueberry note (with some stone fruit and spice). They are medium to heavy in body and with moderate to good acidity. Although there are many heirloom cultivars in Harar, two common bean types are often recognised: Harar longberry, a coffee with a rather large, elongated bean; and Amber coffee, which has an amber-coloured bean, believed to be caused by specific soil conditions. Harar coffee is much sought-after but there are often issues with quality. There is a long-standing tradition in Harar of drinking tea made from the leaves of coffee, and also of using the fleshy part of the coffee fruit to make hot and cold beverages.

There has been a decline in the production of Harar coffee, possibly as a result of climate-induced stress, although conversion to production of the narcotic khat (*Catha edulis*) has probably had a stronger influence on this decline. Since at least the 1960s, farmers have moved to growing this more lucrative crop at the expense of coffee farming and best-practice coffee agronomy.

Harar is divided into three coffee areas, from west to east: Arsi, West Hararge, and East Hararge. Under the ECX contract system, coffees from these areas are classified as Harar A, B and C, depending on their origin and whether they belong to either speciality or commercial grades. Dire Dawa, a large commercial town, is the ECX delivery centre (East Hararge), and a key node for the transportation of Ethiopian coffee to Djibouti.

C2.1 Arsi (Atlas map page 21)

The Arsi (Harar C) coffee area starts at the base of the escarpment at its most western and southern edge, around Chole (south-east of Adama (Nazret)). Towns include Gololcha, and Mechara, with coffee grown at elevations of 1,500–2,000 m. The more successful coffee farms are located above 1,700 m, and use terracing and some shade provided by indigenous trees. In the southern part of Arsi, there is a large shade coffee plantation at Gololcha, which is partly irrigated.

C2.2 West Hararge (Atlas map pages 21, 22)

The West Hararge (Harar B and A) coffee area forms the middle section of the Harar coffee zone, starting north of Mechara and continuing through the towns of Gelemso, Asbe Teferi, Hirna and slightly beyond. Coffee is grown here at elevations of 1,700–2,200 m.

C2.3 East Hararge (Atlas map pages 22, 23, 30, 31)

The East Hararge (Harar A and B) coffee area starts around Kobo and Deder and reaches to the city of Harar, with the chartered city of Dire Dawa to the north. The elevation for coffee ranges from 1,900–2,200 m. On travelling through this coffee area, it appears that very little coffee is being grown. Much of the coffee is confined to suitable microclimates or where the water table is close to the surface, or where water is available for irrigation. There is also production south of Harawacha and around the towns of Bedeno and Girawa.

11. References

1. Davis, A. P., Gole, T. W., Baena, S. & Moat, J. (2012). The impact of climate change on natural populations of Arabica coffee: predicting future trends and identifying priorities. *PLoS ONE* 7: e47981.
2. Minten, B., Tamru, S., Kuma, T. & Nyarko, Y. (2014). Structure and Performance of Ethiopia's Coffee Export Sector. Working paper 66. June 2014. Addis Ababa: Ethiopian Development Research Institute (EDRI) and International Food Policy Research Institute (IFPRI). 30 pages.
3. Tefera, A. (2015). Ethiopia: Coffee Annual Report. GAIN Report (number ET1514–26/5/2015). USDA Foreign Agricultural Service. 6 pages.
4. Tefera, A. & Tefera, T. (2014). Ethiopia: Coffee Annual Report. GAIN Report (number ET1402–13/5/2014). USDA Foreign Agricultural Service. 11 pages.
5. Friis, I., Demissew, S. & Breugel, P. V. (2010). Atlas of the potential vegetation of Ethiopia. *Biol. Skrif.* 58: 1–307.
6. United States Geological Survey (2015). Landsat Missions. Landsat 8. Website: http://landsat.usgs.gov/landsat8.php. Data http://earthexplorer.usgs.gov/ [Accessed Feb 2014].
7. Matthew, M. W., Adler-Golden, S. M., Berk, A., Richtsmeier, S. C., Levine, R. Y., Bernstein, L. S., Acharya, P. K., Anderson, G. P., Felde, G. W., Hoke, M. P., Ratkowski, A., Burke, H.-H., Kaiser, R. D. & Miller, D. P. (2000). Status of atmospheric correction using a MODTRAN4-based algorithm. In: *Proc. SPIE 4049 Algorithms for Multispectral, Hyperspectral, and Ultraspectral Imagery* VI: 199–207.
8. Zhu, Z., Wang, S. & Woodcock, C. E. (2015). Improvement and expansion of the Fmask algorithm: cloud, cloud shadow, and snow detection for Landsats 4–7, 8, and Sentinel 2 images. *Remote Sens. Environ.* 159: 269–277.
9. Kriegler, F. J., Malila, W. A., Nalepka, R. F. & Richardson, W. (1969). Preprocessing transformations and their effects on multispectral recognition. Proceedings of the Sixth International Symposium on Remote Sensing of Environment, Volume II: 97–131.
10. Moat, J., Williams, J., Baena, S., Wilkinson, T., Gole, T. W., Challa, Z. K., Demissew, S. & Davis, A. P. (2017a). Resilience potential of the Ethiopian coffee sector under climate change. *Nat. Plants* 3: art. no. 17081.
11. Hijmans, R. J., Cameron, S. E., Parra, J. L., Jones, P. G. & Jarvis, A. (2005). Very high resolution interpolated climate surfaces for global land areas. *Int. J. Climatol.* 25: 1965–1978.
12. Thuiller, W., Georges, D. & Engler, R. (2014). Biomod2: Ensemble platform for species distribution modeling. R package version 3.1-64. http://CRAN.R-project.org/package=biomod2.
13. Kelbessa, E. & Demissew, S. (2014). Diversity of vascular plant taxa of the flora of Ethiopia and Eritrea. *Ethiopian J. Sci.* 13 (Supp.): 37–45.
14. Tadesse, M., Alemu, B., Bekele, G., Tebikew, T., Chamberlin, J. & Benson, T. (2006). Atlas of the Ethiopian Rural Economy. Central Statistical Agency (Addis Ababa, Ethiopia), International Food Policy Research Insitute (IFPRI) and Ethiopian Development Research Institute (Addis Ababa, Ethiopia). Pp. 93. http://www.ifpri.org/publication/atlas-ethiopian-rural-economy.
15. Tadesse, M. (2017). *Ethiopia — Home of Arabica Coffee: Early Use, Folklore, Coffee Ceremony, Origin and Biology.* CreateSpace Independent Publishing Platform (North Charleston, South Carolina). 110 pages.
16. Wellman, F. L. (1961). *Coffee: Botany, Cultivation and Utilisation*. Leonard Hill, London, UK; Interscience, New York, USA. 488 pages.
17. Wrigley, G. (1988). *Coffee — Tropical Agriculture Series*. Longman Scientific & Technical, Harlow, UK. 639 pages
18. Africa Rikai Project (2012). *The Legend of Ethiopian Coffee — Coffee from its Birthplace*. Gudina Tumsa Publishing House, Addis Ababa, Ethiopia. 53 pages.
19. Ukers, W. H. (1922). *All About Coffee*. Tea and Coffee Trade Journal, New York, USA. 796 pages.
20. Davis, A. P., Govaerts, R., Bridson, D. M. & Stoffelen, P. (2006). An annotated taxonomic conspectus of the genus *Coffea* (Rubiaceae). *Bot. J. Linn. Soc.* 152: 465–512.
21. Tesfaye, K., Borsch, T., Govers, K. & Bekele, E. (2007). Characterisation of *Coffea* chloroplast microsatellites and evidence for the recent divergence of *C. arabica* and *C. eugneioides* chloroplast genomes. *Genome* 50: 1112–1129.
22. Davis, A. P., Tosh, J., Ruch, N. & Fay, M. (2011). Growing coffee: *Psilanthus* (Rubiaceae) subsumed on the basis of molecular and morphological data; implications for the size, morphology, distribution and evolutionary history of *Coffea*. *Bot. J. Linn. Soc.* 167: 357–377.
23. Senbeta, F., Denich, M., Böhmer, H. J., Gole, T. W., Teketay, D. & Demissew, S. (2007). Wild *Coffea arabica* L. in the Afromontane rainforests of Ethiopia: distribution, ecology and conservation. SINET: *Ethiop. J. Sci.* 30: 13–24.
24. Mittermeier, R. A., Robles-Gil, P., Hoffmann, M., Pilgrim, J., Brooks, T., Mittermeier, C. G., Lamoreux, J. & Da Fonseca, G. A. B. (2004). *Hotspots Revisited: Earth's Biologically Richest and Most Endangered Terrestrial Ecoregions.* CEMEX (Mexico City). 392 pages.
25. Meyer, F. G. (1965). Notes on wild *Coffea arabica* from southwestern Ethiopia, with some historical considerations. *Econ. Bot.* 19: 136–151.

26. Sylvain, P. G. (1958). Ethiopian coffee: its significance to world coffee problems. *Econ. Bot.* 12: 111–139.

27. Nutman, F. J. (1933a). The roots system of *Coffea arabica* L. I: Root systems in typical soils of British East Africa. *Empire J. Exp. Agric.* 1: 271–284.

28. Nutman, F. J. (1933b). The roots system of *Coffea arabica* L. II: The effects of some soil conditions in modifying the 'normal' root system. *Empire J. Exp. Agric.* 1: 285–296.

29. Nutman, F. J. (1934). The roots system of *Coffea arabica* L. III: The spatial distribution of the absorbing area of the root. *Empire J. Exp. Agric.* 2: 293–302.

30. DaMatta, F. M., Ronchi, C. P., Maestri, M. & Barros, R. S. (2007). Ecophysiology of coffee growth and production. *Braz. J. Plant Physiol.* 19: 485–510.

31. Hailu, B., Maeda, E. E., Heiskanen, J. & Pellikka, P. (2015). Reconstructing pre-agricultural expansion vegetation cover of Ethiopia. *Appl. Geogr.* 62: 357–365.

32. Moat, J., Williams, J., Baena, S., Wilkinson, T., Demissew, S., Challa, Z. K., Gole, T. W. & Davis, A. P. (2017b). *Coffee Farming and Climate Change in Ethiopia: Impacts, Forecasts, Resilience and Opportunities — Summary Report 2017*. The Strategic Institutions Programme (SCIP). Royal Botanic Gardens, Kew. 37 pages.

33. Hylander, K., Nemomissa, S., Delrue, J. & Enkosa, W. (2013). Effects of coffee management on deforestation rates and forest integrity. *Conserv. Biol.* 27: 1031–1040.

34. EFAP. (1994). *Ethiopian Forest Action Program.* Final report. EFAP Secretariat. Addis Ababa, Ethiopia. 22 pages.

35. Tadesse, D., Gole, T. W. & Senbeta, F. (2008). Deforestation of Afromontane rainforests in Ethiopia. In: Sanchez, I. B. & Alonso, C. L. (eds). *Deforestation Research Progress*, Pp. 19–39. Nova Science Publishers, Inc., New York, USA.

36. Hansen, M. C., Potapov, P. V., Moore, R., Hancher, M., Turubanova, S. A., Tyukavina, A., Thau, D., Stehman, S. V., Goetz, S.J., Loveland, T. R., Kommareddy, A., Egorov, A., Chini, L., Justice, C. O. & Townshend, J. R. G. (2013). High-resolution global maps of 21st century forest cover change. *Science* 342: 850–53. Data available online from: http://earthenginepartners.appspot.com/science-2013-global-forest

37. Gole, T. W. & Davis, A. P. (2013). Coffee as a sustainable forest product. In: *Coffee. A Global Success.* Lack, H. W., Grotz, K. & T. W. Gole (eds). BGBM, Berlin, Germany. Pp. 76–79.

38. Gole, T. W., Itana, A., Tsegaye, B. & Senbeta, F. (2016). *Coffee, Ethiopia's Gift to the World: the Traditional Production Systems as Living Examples of Crop Domestication and Sustainable Production, and an Assessment of Different Certification Schemes.* Environment and Coffee Forest Forum/Institute for Sustainable Development, Addis Ababa, Ethiopia. 51 pages.

39. International Coffee Organisation (ICO) (2017). Historical Data on the Global Coffee Trade. Website: http://www.ico.org/new_historical.asp?section=Statistics

40. Gilbert, S. M. (1938). Variability in yield of *Coffea arabica*. *E. Afr. Agric. J.* 3: 131–139.

41. Netsere, A. (2015). Advance in Arabica forest coffee management research in Ethiopia. *J. Biol. Agric. Healthcare* 5: 31–35.

42. Netsere, A., Kufa, T. & Shimber, T. (2015). Review of Arabica coffee management research in Ethiopia. *J. Biol. Agric. Healthcare* 5: 235–258.

43. Gaitán, A. L., Cristancho, M. A., Castro Caicedo B. L., Rivillas, C. A. & Cadena Gómez, G. (2015). *Compendium of Coffee Diseases and Pests.* APS Press (St. Paul, USA). 79 pages.

44. Jaramillo, J., Muchugu, E., Vega, F. E., Davis, A. P., Borgemeister, C. & Chabi-Olaye, A. (2011). Some like it hot: The influence and implications of climate change on coffee berry borer (*Hypothenemus hampei*) and coffee production in East Africa. *PLoS ONE* 6: e24528.

45. Taye, K., Ashenafi, A., Alemseged, Y., Teshome, K. & Wondiyfraw, T. (2011). The contribution of coffee research for coffee seed development in Ethiopia. *J. Agr. Res. Dev.* 1: 9–16.

46. Yonas, B., Bayetta, B. & Chemeda, F. (2014). Performance evaluation of indigenous Arabica coffee genotypes across different environments. *J. Plant. Breed. Crop. Sci.* 6: 171–178.

47. Tesfaye, K., Borsch, T., Govers, K. & Bekele, E. (2014). ISSR fingerprinting of *Coffea arabica* throughout Ethiopia reveals high variability in wild populations and distinguishes them from landraces. *Plant. Syst. Evol.* 300: 881–897.

48. Tadele, S., Mekbib, F. & Tesfaye, K. (2014). Genetic diversity of coffee (*Coffea arabica* L.) landraces from southern Ethiopia as revealed by inter simple sequence repeat marker. *J. Agr. Sci.* 3: 24–34.

49. Boot, W. J. (2011). *Ethiopian Coffee Buying Manual. Practical Guidelines for Purchasing and Importing Ethiopian Speciality Coffee Beans.* Fintac Inc. for USAID/Ethiopia's Agribusiness and Trade Expansion, Addis Ababa, Ethiopia. 82 pages. https://bootcoffee.com/wp-content/uploads/2013/01/Ethiopian_Coffee_Buying_Guide.pdf

50. ECX Coffee Contracts (2015). Ethiopian Commodity Exchange (ECX)Pp. 14. http://www.ecx.com.et/downloads/contracts/Coffee/CoffeeContracts.pdf

51. Jarvis, A., Reuter, H. I., Nelson, A. & Guevara, E. (2008). Hole-filled SRTM for the globe Version 4, available from the CGIAR-CSI SRTM 90m Database (http://srtm.csi.cgiar.org).

12. Scope and use of Atlas

Coffee suitability

Two main categories are used in the Atlas for coffee (niche) suitability: Forest and Non-Forest. 'Forest' covers areas where there is indigenous forest or significant agroforestry present, and is divided into three colours depending on the suitability for coffee farming: Fair (green-grey), Good (light green) and Excellent (dark green). 'Non-forest' refers to areas where there is little or no forest but there is suitability for coffee farming: Fair (pink), Good (blue-grey) and Excellent (mauve). In most cases, shaded areas for forest (or significant agroforestry) indicate actual coffee-growing areas, although the intensity of farming will vary. Areas shaded for non-forest could grow coffee with suitable intervention (e.g. shade or improved shade cover from the planting of suitable indigenous shade trees) depending on the suitability category (Fair, Good or Excellent). Or, there may be non-forest coffee being farmed in these areas. The three coffee farming suitability categories are explained below. See also Key to Symbols on Atlas map page 1.

Fair. Coffee farming is possible and sometimes good, but may require additional inputs (e.g. irrigation, shade or improved shade). Yields poor in years with low rainfall and high temperatures.

Good. Coffee farming good, with the potential for consistent, high-quality yields. Yield and quality may be negatively influenced in some years (by adverse weather conditions, especially low rainfall and high temperatures).

Excellent. Coffee farming good to excellent, with the potential for consistent, high-quality yields. Yields and quality can be negatively influenced by adverse weather conditions, but this is less likely than in other areas.

Figure 6 (page 29) provides summarised information for coffee farming suitability and forest cover; the Atlas maps show the coffee farming suitability in detail.

Wild coffee forests

Wild Arabica coffee is found in nearly all the forested areas with Excellent or Good coffee (niche) suitability, which is marked dark green on the Atlas maps, with the exception of most parts of the western Sidamo coffee area. Wild coffee forests occur in the western part of A2.1 Wellega (Kelem Wellega), A2.2 Illubabor, A2.3 Jimma-Limu, A2.4 Tepi, A2.5 Kaffa, A2.6 Bench Maji, and C1.2 Bale coffee areas (see Chapter 10 and Atlas map pages).

Other (non-coffee) humid forest

The grey shading on the maps mainly denotes humid forest of the Moist evergreen Afromontane forest (MAF) and Transitional rain forest (TRF)[5] types, but lacking (or currently lacking) suitability for coffee. Further information on these forest types is provided in Chapter 8 (Coffee farming). Humid forest occurs at elevations between 450 m and c. 3,000 m, and so can occur outside of the current elevation range for coffee suitability (900–2,300 m). Humid forest is now rare in Ethiopia below 900 m; lower elevation humid forest (450–900 m) is mainly restricted to western Ethiopia in the far western part of the Ethiopian coffee landscape (see Atlas map pages 8, 9, 16, 17). Higher elevation humid forest areas (e.g. above 2,300 m) may come into suitability as the climate warms[10].

Coffee gardens (coffee bean symbol)

The scale of coffee-growing in the coffee garden (garden coffee (GC)) areas ranges from a few trees to small plots (e.g. 10–50 coffee trees), to medium-sized plots (e.g. 100–500 trees). Yields may be poor and inconsistent, or good, largely depending on inputs (e.g. irrigation) and/or local conditions (e.g. microclimate or water table). Coffee gardens are mostly found in full sun or light shade, and often close to households.

Other coffee information

Coffee zones and areas. Coffee zones and coffee areas are listed, mapped, and described in Chapter 10 (The Coffee Areas).

Coffee localities. The symbol (an upright Ethiopian coffee pot (jebena)) denotes a significant or valuable coffee-growing locality (origin), associated with the coffee that is growing there (e.g. Limu, Kaffa, Bale). Some lesser-known localities have been included to better detail the coffee farming landscape, including those at the extremities (e.g. Konso, Zege) and in isolated places (e.g. Debre Zeyit) within the coffee-growing landscape of Ethiopia.

Coffee towns. The symbol (a tilted, pouring, Ethiopian coffee pot (jebena)) is used in a similar way as for coffee localities, but in this case the coffee locality is directly associated with the town's name (e.g. Agaro, Jimma, Bonga, Yirgacheffe). In fact, in some coffee towns, a large tilting coffee pot (jebena) is placed in the centre of town (e.g. in Jimma and Bonga).

Coffee delivery centres. The symbol of a factory building denotes an ECX coffee delivery centre, from where the surrounding region's coffee is delivered for the later stages of processing (e.g. hulling, grading, weighing) and ongoing transportation. The delivery centres represent key nodes in the logistic framework for Ethiopian coffee processing and transportation. Some delivery centres are also coffee towns (see above), for example Bonga and Jimma, whereas others are not (e.g. Dilla).

Other mapping information

Regions, zones, woredas and kebeles

Addis Ababa is the capital city of Ethiopia. Ethiopia is divided into a hierarchy of administrative divisions, including regions, chartered cities, zones, woredas and kebeles. There are nine regions (capital towns are given in parentheses): Afar (Semera), Amhara (Bahir Dar), Benishangul Gumuz (Asosa), Gambella (Gambela), Harari (Harar), Oromia (Addis Ababa/Finfinne), Somali (Jijiga/Jigjiga), Southern Nations, Nationalities, and Peoples' Region [SNNPR] (Hawassa) and Tigray (Mekele).

Addis Ababa and Dire Dawa are the two chartered cities. The regions are subdivided into zones (second-level division), and are displayed in Figure 8 and on Atlas map page 1. Woredas (third-level subdivision) are generally subdivisions of zones, and each has its own elected council. Kebeles (fourth-level subdivision) are the smallest administrative unit and represent an administrative delimitation at the local level. The Atlas does not provide information at the level of Woreda and Kebele.

The names of zones, regions, woredas and kebeles are subject to change and regional variation, and it should be noted that those used in the Atlas are unlikely to precisely follow the most recent nomenclature and geographical boundaries.

Settlements, transport and navigation

Inevitably there will be some inaccuracies in the Atlas, due to outdated mapping data, especially for roads, settlements (regional capitals, towns and larger villages) and railways. In particular, many roads are being upgraded and new ones built (see below). For these reasons, the mapping data presented in the Atlas should not be relied upon for detailed, accurate navigation. A navigation disclaimer is given on page 3.

Settlements have been classified into three categories on the Atlas map pages: (1) regional capitals (upper case; e.g. GAMBELA); (2) major towns (bold; e.g. **Arba Minch**); and (3) other towns (non-bold; e.g. Alem Teferi). The last two of these categories (major towns and other towns) provide the user of the Atlas with an indication of size and amenities. Thus, for the traveller, major towns will normally have several restaurants, accommodation (hotel or other), at least one bank, and in many cases, at least one fuel/service station. Some of the larger towns will have an airport or airstrip (see Gazetteer). Major amenities are present in regional capitals.

We have done our best to update and classify the roads, but the Atlas cannot be relied upon for either accuracy or up-to-date information. Ethiopian roads are in a continual state of development, with upgrades, new routes, and some minor or little-used roads falling into disrepair. If in doubt it is always a good idea to ask locally for information on road condition and specific routes.

The Atlas includes the recently completed electrified railway, which runs from Addis Ababa to Djibouti.

A gazetteer (geographical name index) is provided on Atlas map pages 43–59, to enable the user to find zones, coffee-growing localities, coffee towns, coffee delivery centres, regiona capitals, airports, protected areas (e.g. protected areas and wildlife reserves), major lakes, summits/mountains and towns/ settlements. The names of settlements (e.g. towns) are subjec to change and regional variation.

Elevation and topology

The letter 'm', given in text and on maps, indicates meters and with the associated number (e.g. 1,200 m) refers to the elevation above sea level. In order to portray the elevation and landscape encompassed by the Atlas, the Shuttle Radar Topography Mission (SRTM) digital elevation database[51] has been processed to produce a colour-shaded relief image map. The colouring corresponds directly to elevation, which when combined with the relief shading shows the topology in the Atlas map pages; major peaks (mountains) are named and indicated (with a triangle). Figure 8 represents an elevation map for the whole of Ethiopia, overlaid with the Atlas page index. This will enable to reader to gauge the approximate elevations for a given area or locality.

As a further guide to elevation, the coffee growing areas (i.e. areas coloured to indicate coffee suitability and coffee gardens) mostly range between 1,000–2000 m. It should be noted that Ethiopia is a high elevation country, with most of its land mass above 1,000 m. For further guidance, a spot height is given on a specific peak, on each page (e.g. 2275 m).

View across the Simien Mountains (North Gonder, Amhara)

Image: Jenny Williams

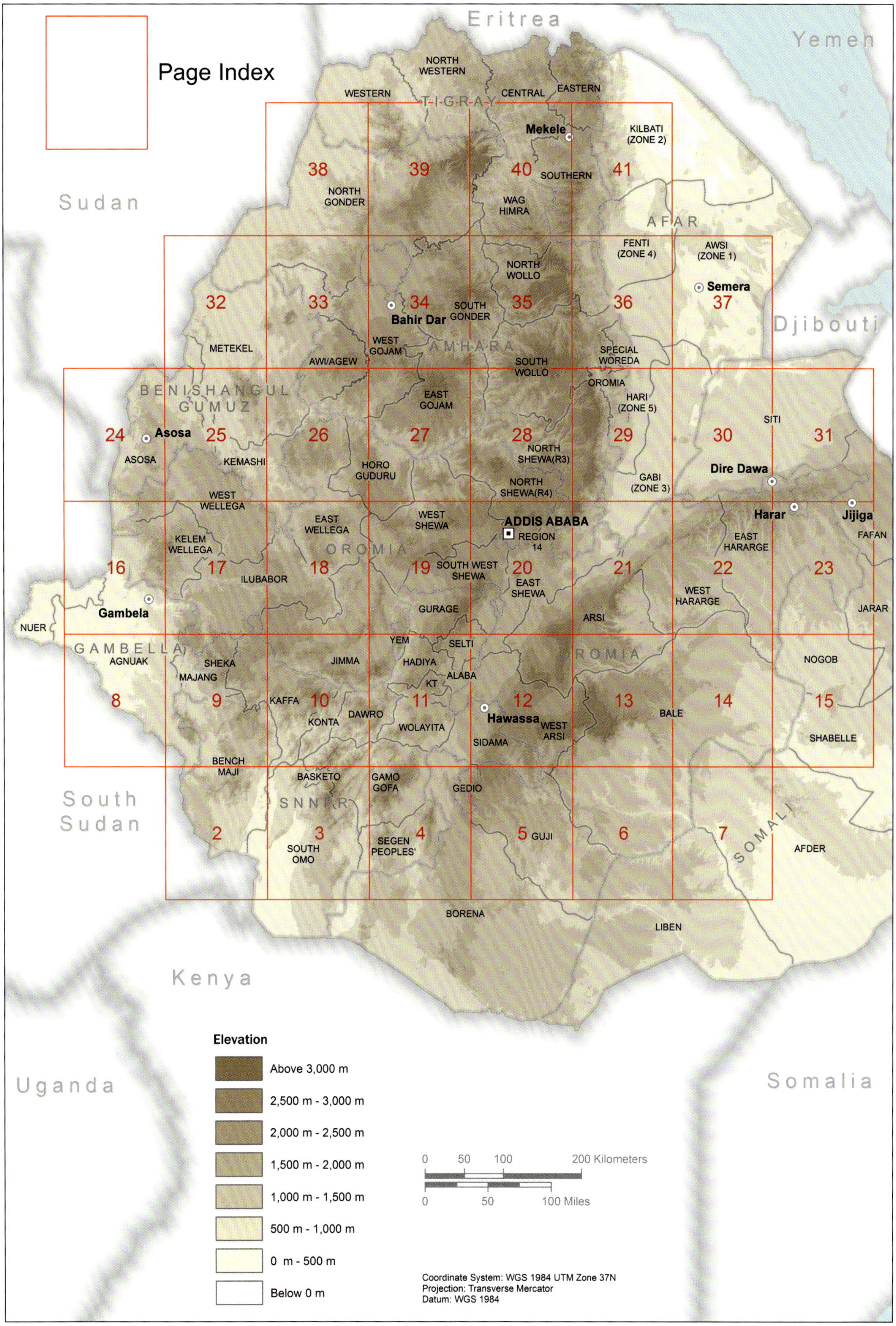

Figure 8. Elevation map. Numbers refer to Atlas map page numbers. Regions are given in large upper-case, zones in smaller upper-case. The capital city (Addis Ababa) is indicated with a square bullet-point, and regional capitals are indicated with round bullet-points. KT = Kembata Tembaro (zone).

Serving traditional Ethiopian coffee
Image: Alan Schaller (courtesy of Union Coffee)

Index

Page numbers in **bold** indicate an image or table.

The Maps

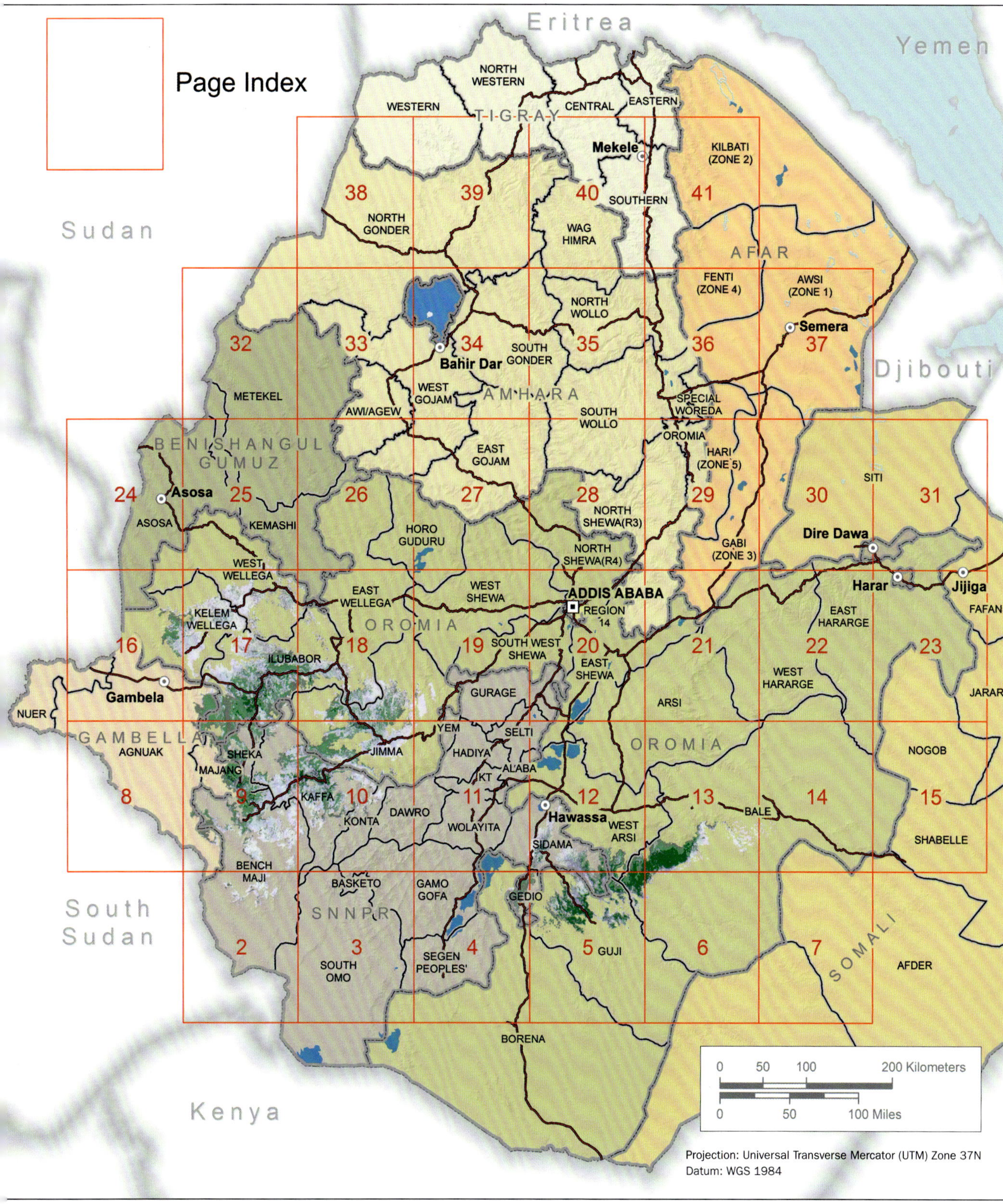

Key to Symbols

PLACES

- Towns
- Airports
- Mountains and Peaks

ROADS AND RAIL

- Main road
- Secondary road
- Other road
- Railway

BORDERS

- State
- Region
- Zone

PROTECTED AREAS

- Protected areas

DRAINAGE

Rivers

- Principal river

Water bodies

- Permanent
- Seasonal

COFFEE SUITABILITY

i. Forest

- Fair
- Good
- Excellent

ii. Non-forest

- Fair
- Good
- Excellent

iii. Other forest

- Humid forest, non coffee

KEY COFFEE-GROWING LOCALITIES

- Principal
- Other

COFFEE TOWNS

- Principal
- Other

COFFEE DELIVERY CENTRE

- Warehouse

COFFEE GARDENS

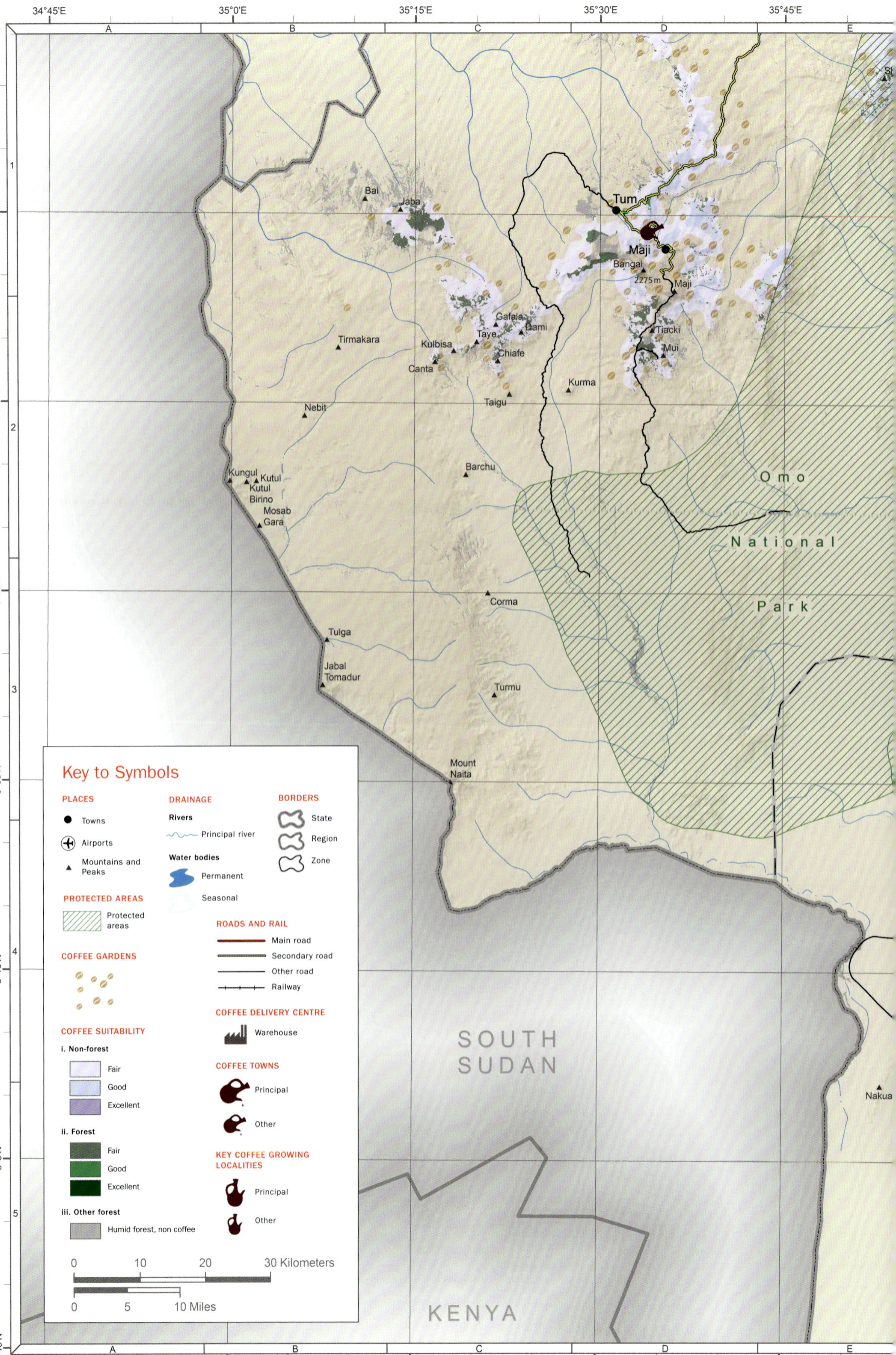

34°45'E
35°0'E
35°15'E
35°30'E
35°45'E
A
B
C
D
E
1
2
3
4
5
6°15'N
6°0'N
5°45'N
5°30'N
5°15'N
5°0'N
4°45'N
Tum
Maji
Bangal
2275 m
Maji
Bai
Jaba
Gafala
Dami
Taye
Kulbisa
Chiafe
Canta
Tiacki
Mui
Tirmakara
Kurma
Taigu
Nebit
Barchu
Kungul
Kutul
Kutul
Birino
Mosab
Gara
Omo
National
Park
Corma
Tulga
Jabal
Tomadur
Turmu
Mount
Naita
SOUTH
SUDAN
KENYA
Nakua
Key to Symbols
PLACES
Towns
Airports
Mountains and Peaks
DRAINAGE
Rivers
Principal river
Water bodies
Permanent
Seasonal
BORDERS
State
Region
Zone
PROTECTED AREAS
Protected areas
ROADS AND RAIL
Main road
Secondary road
Other road
Railway
COFFEE GARDENS
COFFEE DELIVERY CENTRE
Warehouse
COFFEE SUITABILITY
i. Non-forest
Fair
Good
Excellent
ii. Forest
Fair
Good
Excellent
iii. Other forest
Humid forest, non coffee
COFFEE TOWNS
Principal
Other
KEY COFFEE GROWING LOCALITIES
Principal
Other
0
10
20
30 Kilometers
0
5
10 Miles

10

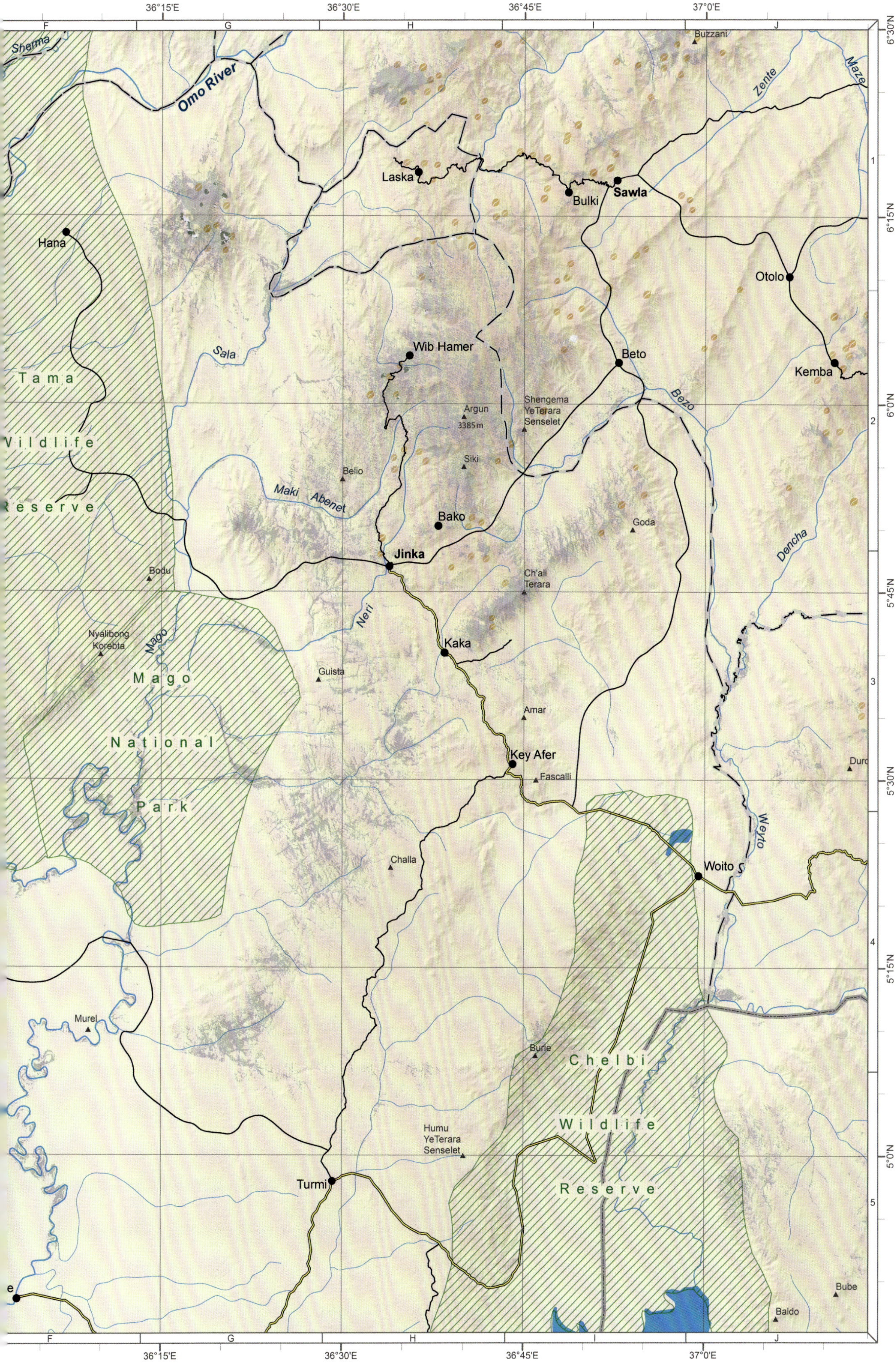

36°15'E
36°30'E
36°45'E
37°0'E
6°30'N
6°15'N
6°0'N
5°45'N
5°30'N
5°15'N
5°0'N
Omo River
Sherma
Zente
Maze
Buzzani
Laska
Bulki
Sawla
Hana
Otolo
Wib Hamer
Beto
Kemba
Sala
Tama
Wildlife
Reserve
Argun
3385 m
Shengema YeTerara Senselet
Bezo
Siki
Belio
Maki
Abenet
Bako
Goda
Dencha
Jinka
Bodu
Ch'ali Terara
Neri
Nyalibong Korebta
Mago
Kaka
Guista
Mago
National
Park
Amar
Key Afer
Fascalli
Duro
Weyto
Challa
Woito
Murel
Burle
Chelbi
Wildlife
Reserve
Humu YeTerara Senselet
Turmi
Bube
Baldo

4

Lake Abaya
Lake Chamo
Nechisar National Park
Zefine
Wajifo
Delle
Selam Ber
Morka
Wacha
Gidicho
Birbir
Darsi
Chencha
Dita
Guge
Gacio
Otolo
Koti
Yirgacheffe
Fiseha Genet
Belta
Bonke Beza
Genta Terara
Kemba
Tempe Terara 3315 m
Arba Minch
Chelelektu
Gede
Geresе
Mele
Tore
Kele
Jelo
Delo
Arguba
Soyema YeTerara Senselet
Mergele Terara
Gidole
Edulei
Melka Jewe
Agere Mariyam
Guroza Terara
Segen
Gato
Duro
Goche
Soyema
Gewada
Gubba
Sillia
Finchaw
Kolme
Bekawile (Konso)
Gemole
Fasha
Yabal
Dedertu
Segen Wenz
Darrito
Teltele
Tedim
Bolekedo
Gundile
Yabelo
Obua
Bube
Calciati
Baldo

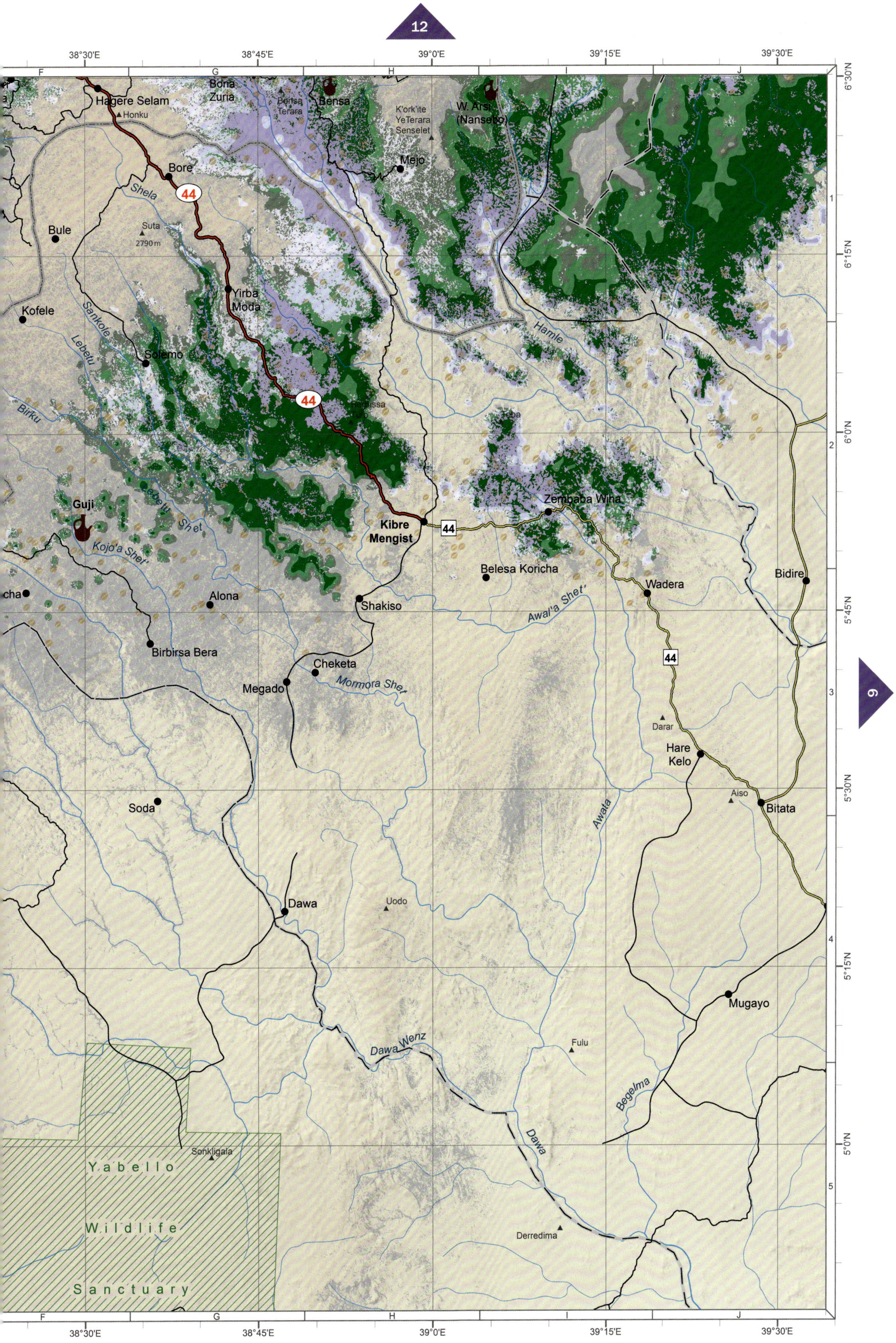
12
6
38°30'E
38°45'E
39°0'E
39°15'E
39°30'E
6°30'N
6°15'N
6°0'N
5°45'N
5°30'N
5°15'N
5°0'N
Hagere Selam
Honku
Bona Zuria
Bensa
K'ork'ite YeTerara Senselet
W. Arsi (Nansebo)
Mejo
Bore
Shela
Suta 2790m
Bule
Yirba Moda
Kofele
Sankole
Lebetu
Solemo
Birku
Hamle
Guji
Kojo'a Shet'
Kibre Mengist
Zembaba Wiha
Belesa Koricha
Bidire
Wadera
Alona
Shakiso
Awal'a Shet'
Birbirsa Bera
Cheketa
Megado
Mormora Shet
Darar
Hare Kelo
Aiso
Bitata
Soda
Awata
Dawa
Uodo
Mugayo
Fulu
Dawa Wenz
Begelma
Dawa
Sonkligala
Yabello Wildlife Sanctuary
Derredima
44

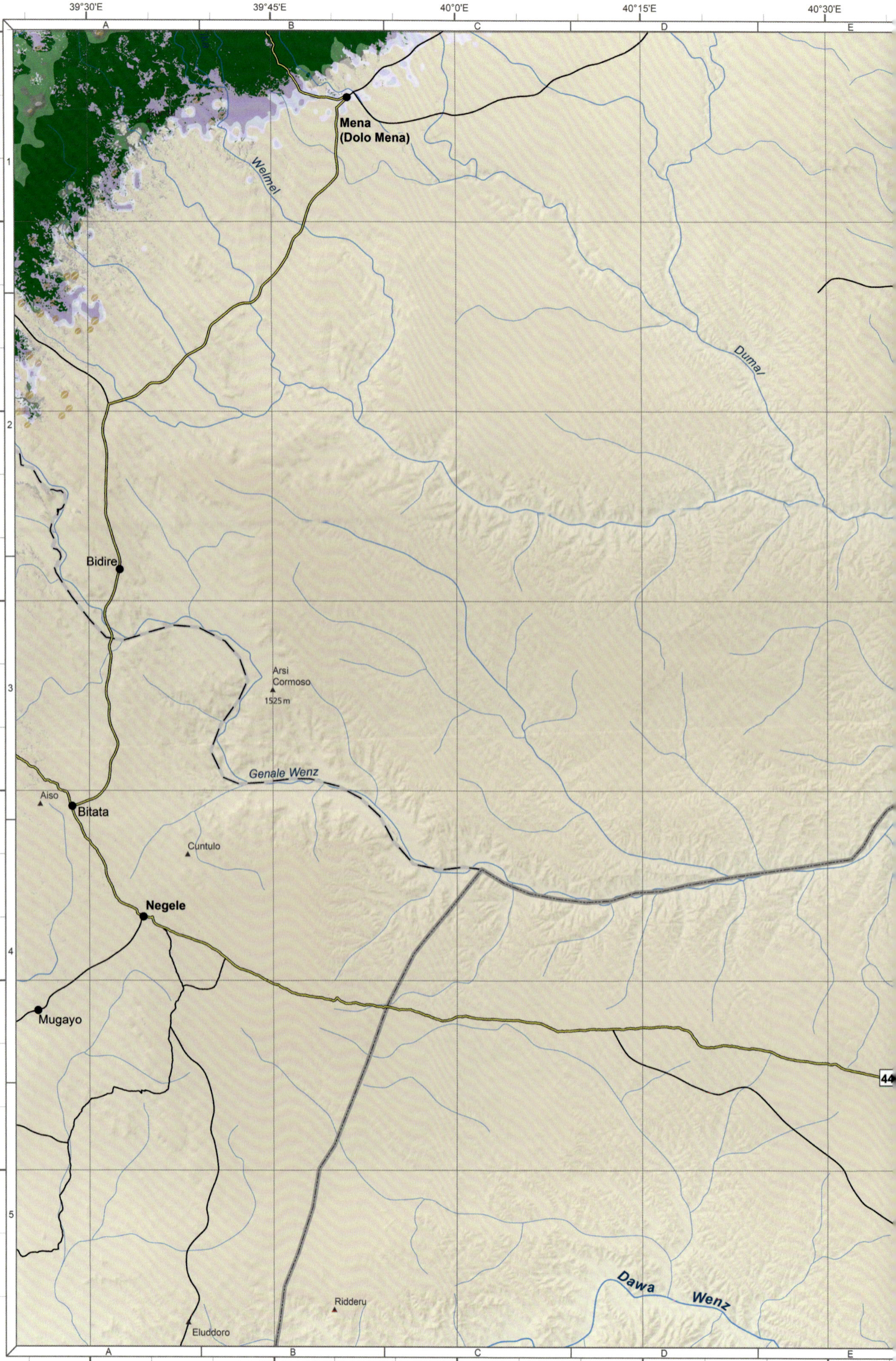

13
5
39°30'E
39°45'E
40°0'E
40°15'E
40°30'E
6°30'N
6°15'N
6°0'N
5°45'N
5°30'N
5°15'N
5°0'N
Mena
(Dolo Mena)
Welmel
Dumal
Bidire
Arsi
Cormoso
1525 m
Genale Wenz
Aiso
Bitata
Cuntulo
Negele
Mugayo
44
Dawa
Wenz
Ridderu
Eluddoro

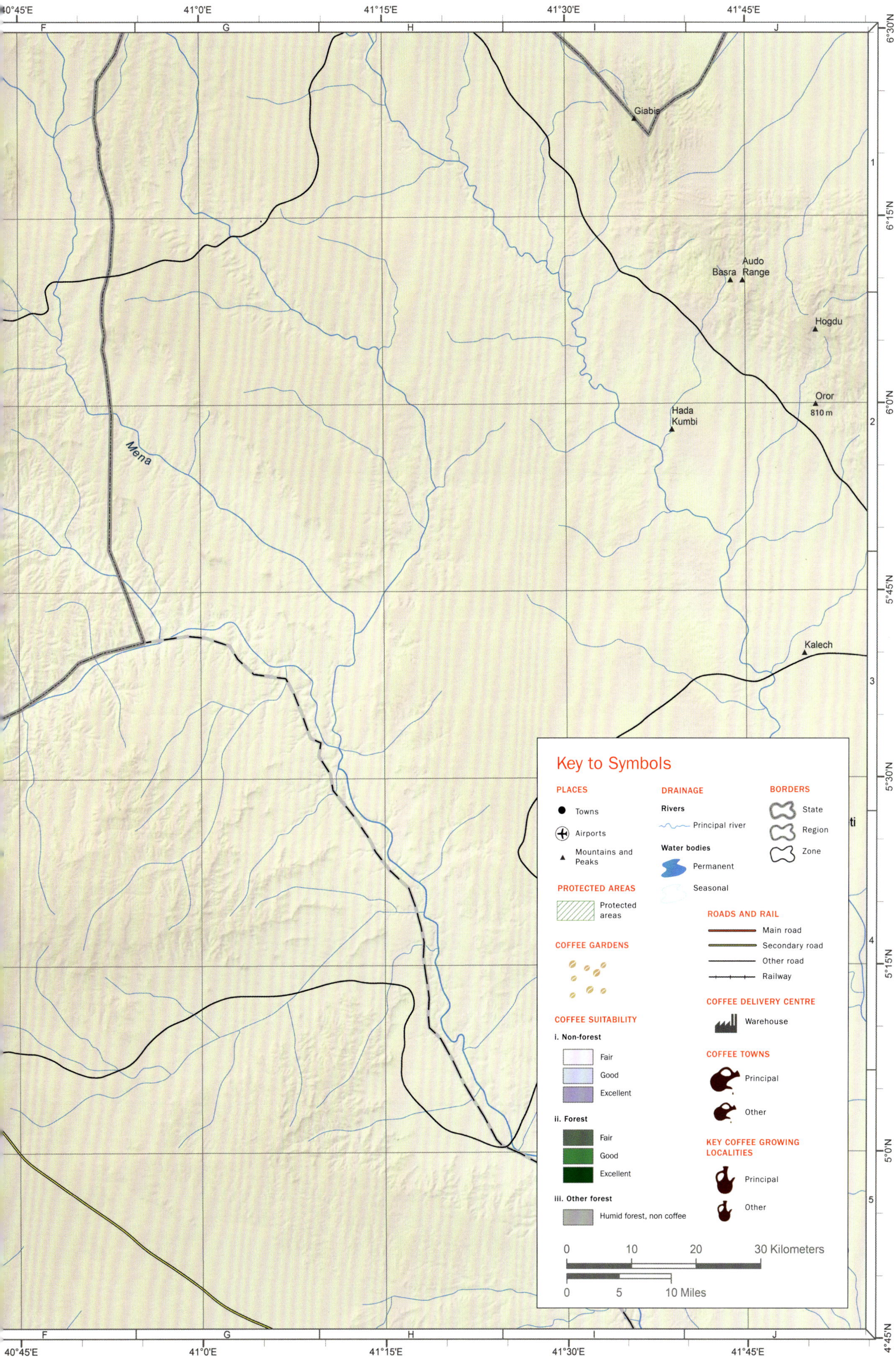

40°45'E
41°0'E
41°15'E
41°30'E
41°45'E
F
G
H
I
J
6°30'N
6°15'N
6°0'N
5°45'N
5°30'N
5°15'N
5°0'N
4°45'N
1
2
3
4
5
Giabis
Audo Range
Basra
Hogdu
Oror
810 m
Hada Kumbi
Mena
Kalech
Key to Symbols
PLACES
Towns
Airports
Mountains and Peaks
DRAINAGE
Rivers
Principal river
Water bodies
Permanent
Seasonal
BORDERS
State
Region
Zone
PROTECTED AREAS
Protected areas
ROADS AND RAIL
Main road
Secondary road
Other road
Railway
COFFEE GARDENS
COFFEE DELIVERY CENTRE
Warehouse
COFFEE SUITABILITY
i. Non-forest
Fair
Good
Excellent
COFFEE TOWNS
Principal
Other
ii. Forest
Fair
Good
Excellent
KEY COFFEE GROWING LOCALITIES
Principal
Other
iii. Other forest
Humid forest, non coffee
0 10 20 30 Kilometers
0 5 10 Miles

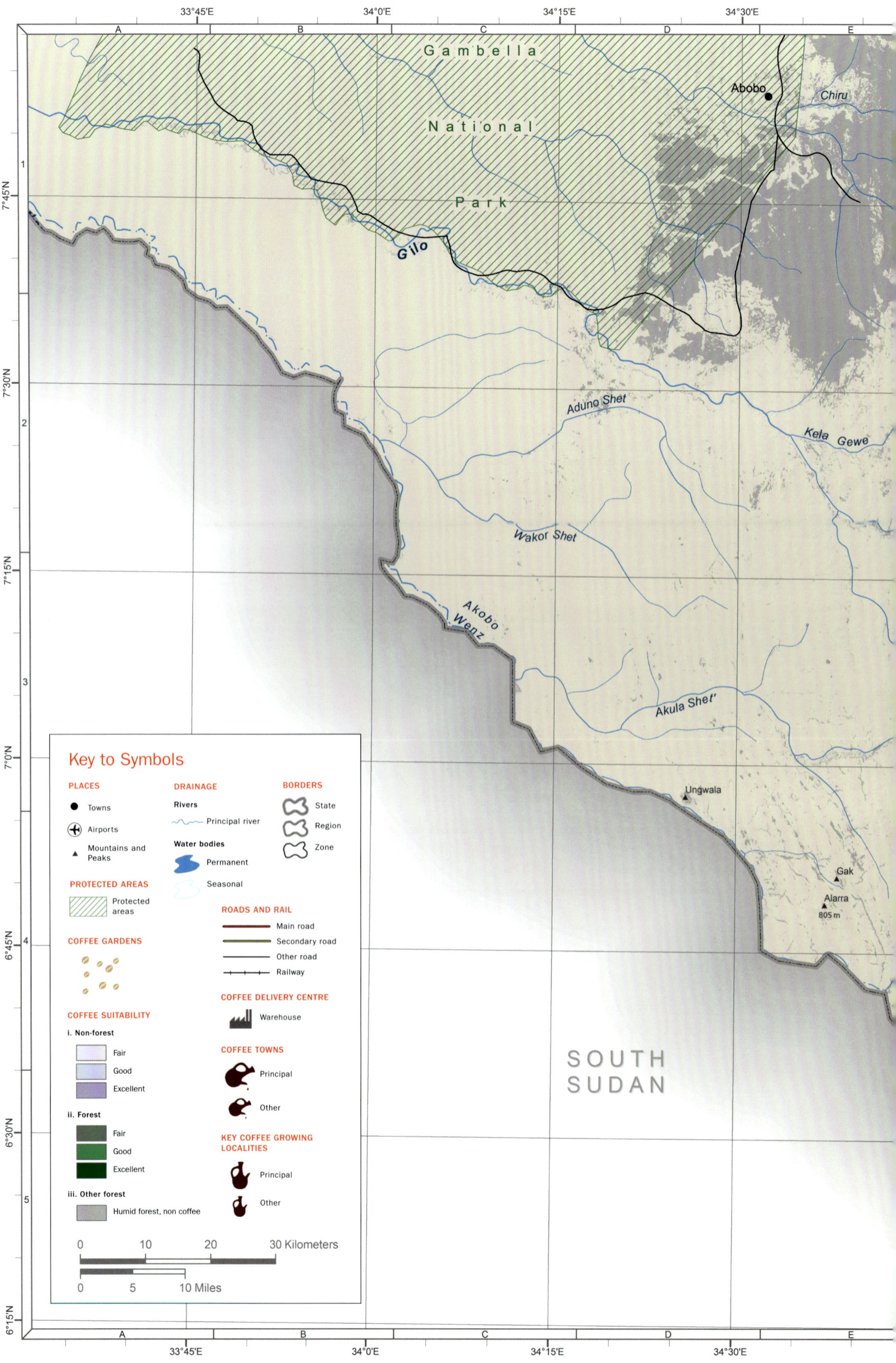
33°45'E
34°0'E
34°15'E
34°30'E
A
B
C
D
E
1
2
3
4
5
7°45'N
7°30'N
7°15'N
7°0'N
6°45'N
6°30'N
6°15'N
Gambella
National
Park
Abobo
Chiru
Gilo
Aduno Shet
Kela Gewe
Wakor Shet
Akobo Wenz
Akula Shet'
Ungwala
Gak
Alarra
805 m
SOUTH SUDAN
Key to Symbols
PLACES
Towns
Airports
Mountains and Peaks
PROTECTED AREAS
Protected areas
COFFEE GARDENS
COFFEE SUITABILITY
i. Non-forest
Fair
Good
Excellent
ii. Forest
Fair
Good
Excellent
iii. Other forest
Humid forest, non coffee
DRAINAGE
Rivers
Principal river
Water bodies
Permanent
Seasonal
BORDERS
State
Region
Zone
ROADS AND RAIL
Main road
Secondary road
Other road
Railway
COFFEE DELIVERY CENTRE
Warehouse
COFFEE TOWNS
Principal
Other
KEY COFFEE GROWING LOCALITIES
Principal
Other
0 10 20 30 Kilometers
0 5 10 Miles

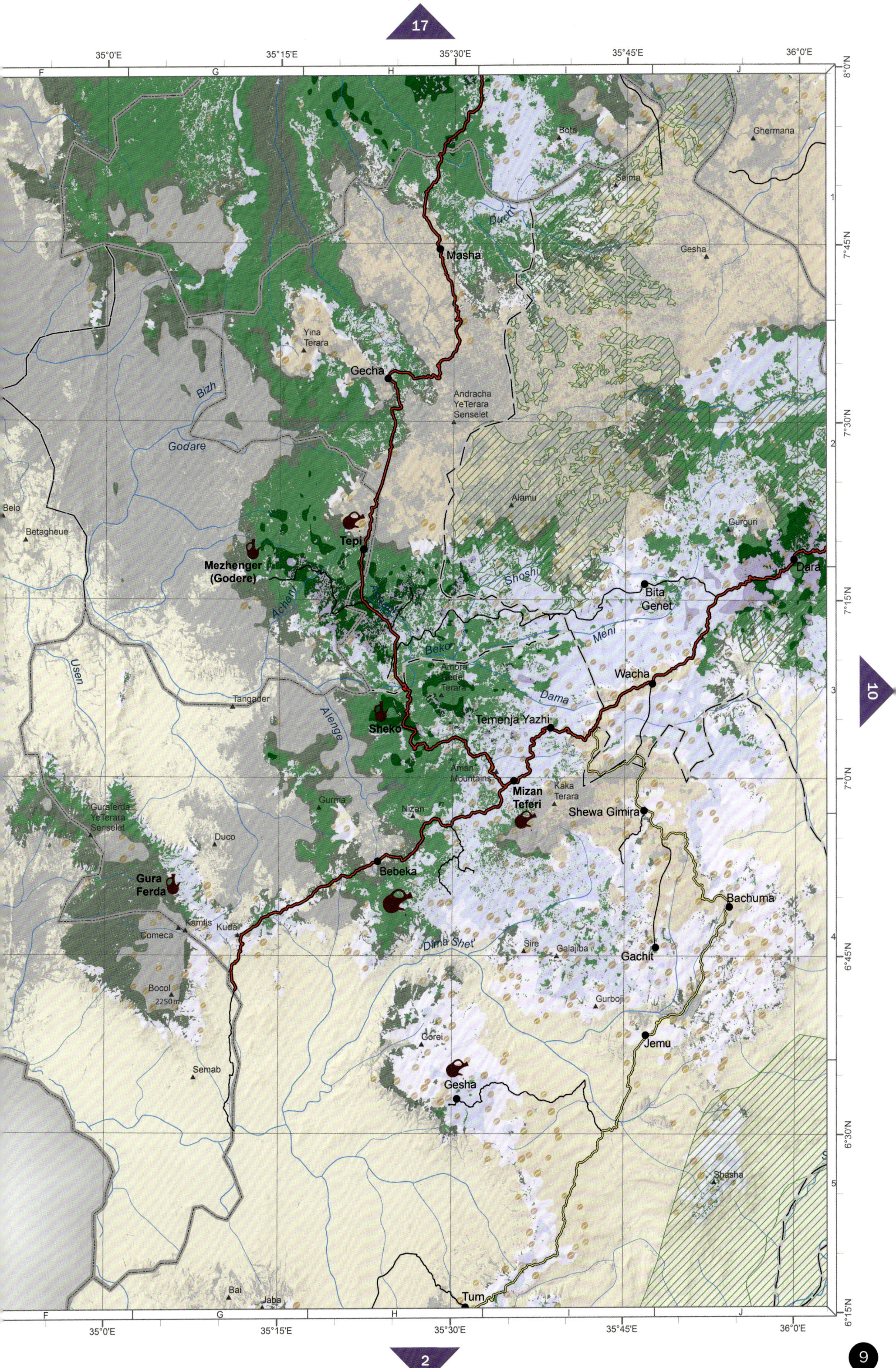
17
10
2
35°0'E
35°15'E
35°30'E
35°45'E
36°0'E
8°0'N
7°45'N
7°30'N
7°15'N
7°0'N
6°45'N
6°30'N
6°15'N
F
G
H
I
J
1
2
3
4
5
Bota
Ghermana
Selma
Dueri
Masha
Gesha
Yina
Terara
Gecha
Bizh
Andracha
YeTerara
Senselet
Godare
Alamu
Belo
Betagheue
Tepi
Gurguri
Mezhenger
(Godere)
Dara
Shoshi
Bita
Genet
Achani
Meni
Usen
Beko
Amora
Gedel
Terara
Wacha
Dama
Tangader
Alenge
Sheko
Temenja Yazhi
Aman
Mountains
Mizan
Teferi
Kaka
Terara
Shewa Gimira
Guraferda
YeTerara
Senselet
Gurma
Nizan
Duco
Bebeka
Gura
Ferda
Bachuma
Kamtis
Kuda
Comeca
Dima Shet
Sire
Galajiba
Gachit
Bocol
2250m
Gurboji
Jemu
Gorei
Semab
Gesha
Shasha
Bai
Jaba
Tum

18

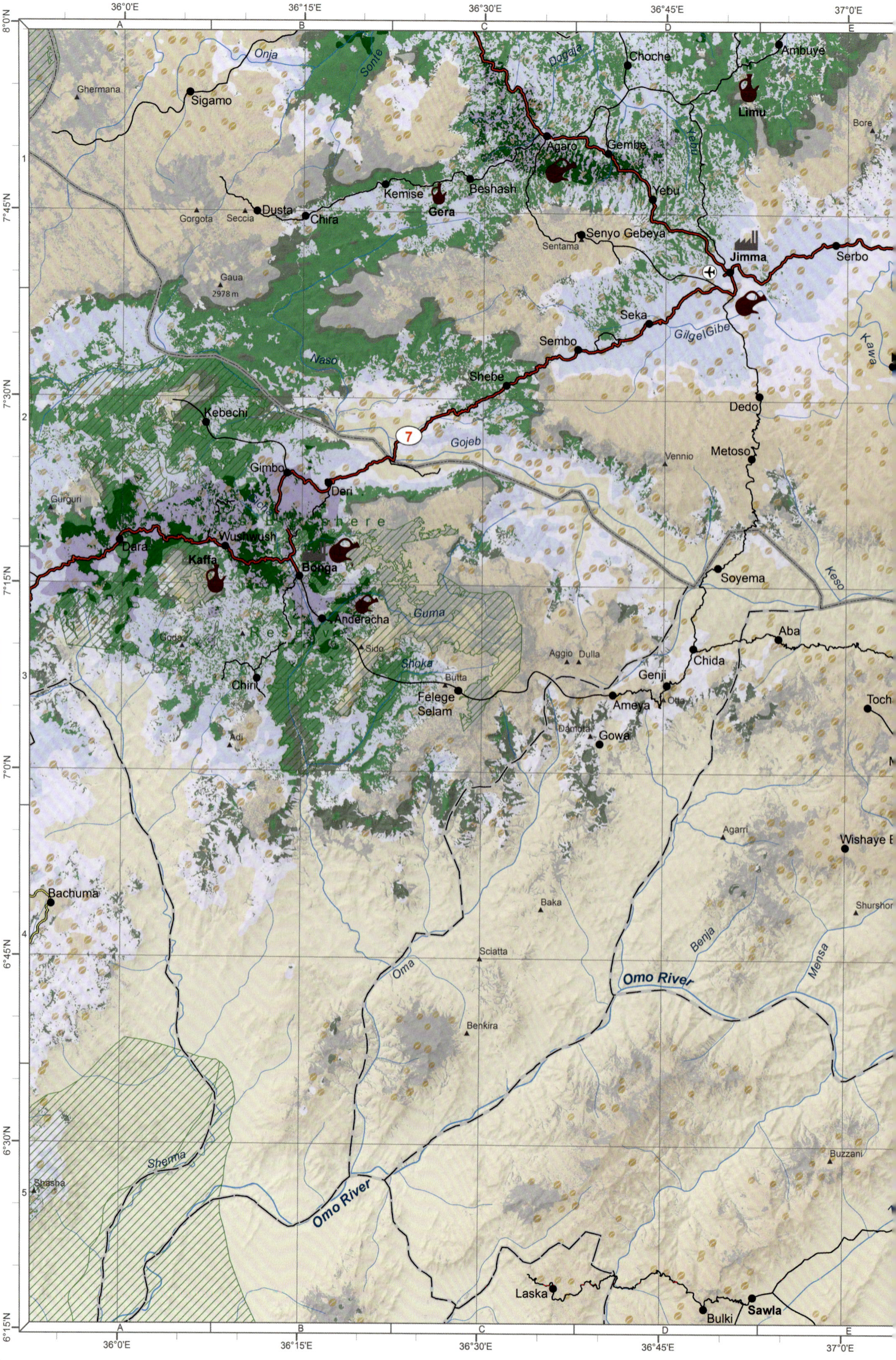
36°0'E
36°15'E
36°30'E
36°45'E
37°0'E
8°0'N
7°45'N
7°30'N
7°15'N
7°0'N
6°45'N
6°30'N
6°15'N
A
B
C
D
E
1
2
3
4
5
Onja
Sonte
Dogaja
Choche
Ambuye
Limu
Bore
Ghermana
Sigamo
Agaro
Gembe
Kemise
Beshash
Gera
Yebu
Gorgota
Seccia
Dusta
Chira
Senyo Gebeya
Sentama
Jimma
Serbo
Gaua
2978 m
Seka
Gilgel Gibe
Sembo
Kawa
Naso
Shebe
Dedo
Kebechi
7
Gojeb
Metoso
Vennio
Gimbo
Deri
Gurguri
Dara
Wushwush
Kaffa
Bonga
Soyema
Keso
Anderacha
Guma
Aba
Goda
Sido
Aggio
Dulla
Chida
Shoka
Butta
Genji
Chiri
Felege
Selam
Ameya
Otta
Toch
Damota
Gowa
Adi
Agarri
Wishaye
Bachuma
Baka
Shurshor
Sciatta
Benja
Oma
Omo River
Mensa
Benkira
Buzzani
Sherma
Shasha
Omo River
Laska
Bulki
Sawla

9

3

19

37°15'E
37°30'E
37°45'E
38°0'E
38°15'E
8°0'N
7°45'N
7°30'N
7°15'N
7°0'N
6°45'N
6°30'N
Saja
Sekoru
Fofa
Deneba
Asendabo
Giangero
Toba
Nedi
Dogosa
Anzeche
Kose
Muggo
Scerit
Lera
Morsito
Boshe
Gambora
Gibe
Tiliku Nada
Nada
Abay Hayk'
Worabe
Dijo
Balch'i
Dalocha
Weyra
Wulaberg
Furfuro
Fonko
9
Hosaina
Alem Gebeya
Jido
Golgio
Guder
Bilate
Mai Gudo
Ajacho
Jajura
Gimbichu
Shenk'ola Terara
Garo
Bobigia
Bubisa
Medula
Sengiya Terara
Gambata
Kulito
Aje
Durgi
Turduna Terara
Ambericho Terara
3020m
Durame
Welde Hane
Angecha
Tambaro
Mazoriya
Cassi
Ropi
Weybo
Shapa
Shone
Areka
Bolosso
Boditi
Damota Terara
Duguna
Bele
Soddo
Dimtu
Bedesa
Koysha Terara
Uolamo
Bisare
Gesuba
Kindo Halato
Umbo
Menisa
Deme
Amesa
Tebela
Aposto
Derba
Chuko
Seware
Bilate
Gedano
Doghe
Uoide
Uoscio
Zefine
Lake Abaya
Tefere Kela
Delle
Selam Ber
Kulano
Wajifo
Morka
Dilla
Gidicho
Shomb YeTerara Senselet
Wacha
Darsi
Birbir
Wenago
Koti
F
G
H
I
J
1
2
3
4
5

12

4

20

11

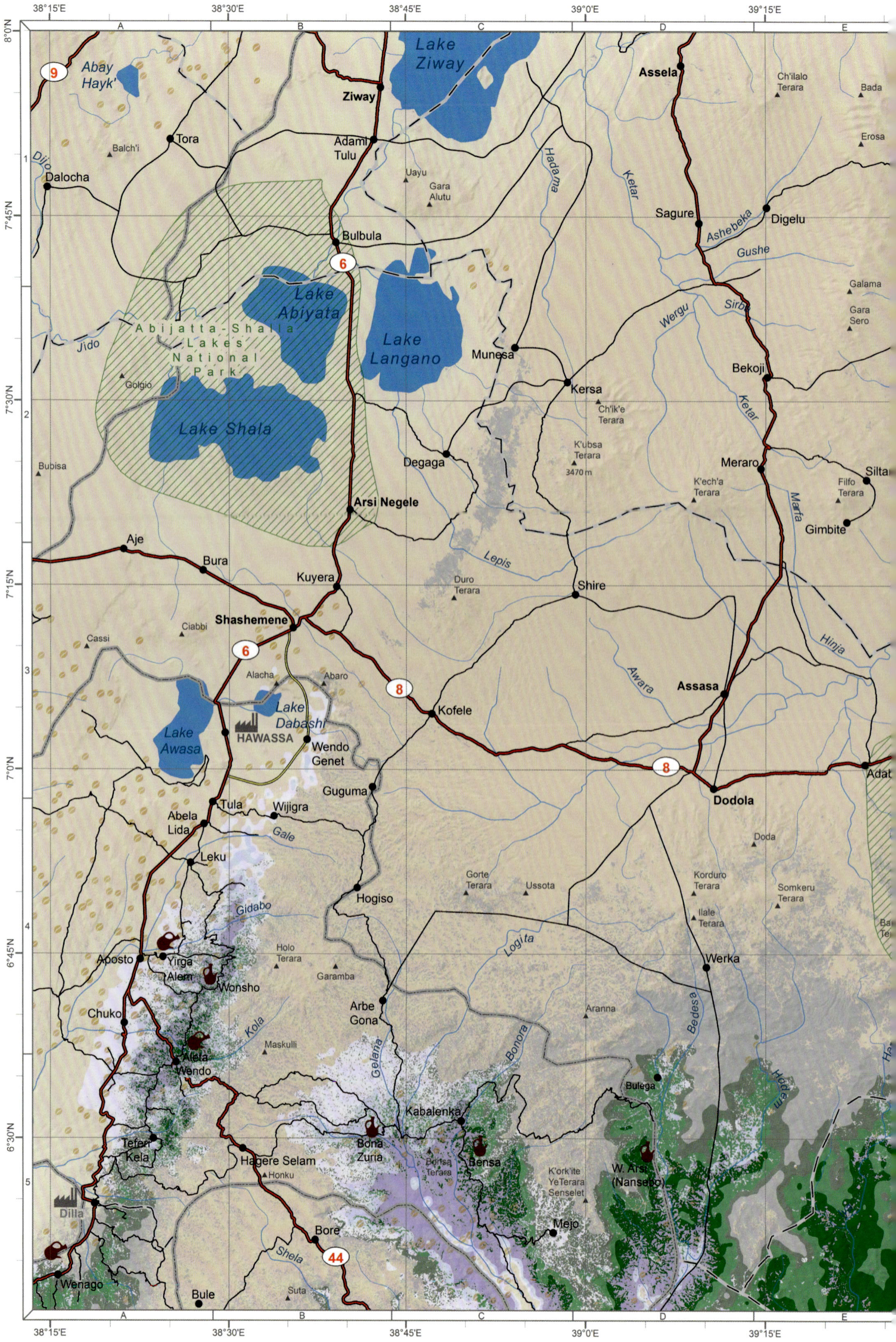

38°15'E
38°30'E
38°45'E
39°0'E
39°15'E
8°0'N
7°45'N
7°30'N
7°15'N
7°0'N
6°45'N
6°30'N
A
B
C
D
E
1
2
3
4
5
Lake Ziway
Abay Hayk'
Assela
Ch'ilalo Terara
Bada
Ziway
Tora
Adami Tulu
Erosa
Balch'i
Dijo
Dalocha
Uayu
Gara Alutu
Hadama
Ketar
Sagure
Digelu
Ashebeka
Bulbula
Gushe
Lake Abiyata
Lake Langano
Abijatta-Shalla Lakes National Park
Jido
Golgio
Galama
Gara Sero
Wergu
Sirba
Munesa
Bekoji
Kersa
Ch'ik'e Terara
Lake Shala
Degaga
K'ubsa Terara
3470 m
Meraro
Silta
Bubisa
K'ech'a Terara
Filfo Terara
Arsi Negele
Marta
Gimbite
Aje
Lepis
Bura
Kuyera
Duro Terara
Shire
Shashemene
Ciabbi
Cassi
Hinja
Alacha
Abaro
Awara
Assasa
Lake Dabashi
Kofele
Lake Awasa
HAWASSA
Wendo Genet
Adat
Dodola
Guguma
Tula
Wijigra
Abela Lida
Gale
Doda
Leku
Gorte Terara
Ussota
Korduro Terara
Somkeru Terara
Hogiso
Ilale Terara
Gidabo
Logita
Holo Terara
Garamba
Aposto
Yirga Alem
Wonsho
Werka
Arbe Gona
Aranna
Chuko
Kola
Bedesa
Maskulli
Bonora
Gelana
Aleta Wendo
Bulega
Kabalenka
Teferi Kela
Hagere Selam
Bona Zuria
Bensa
W. Arsi (Nansebo)
Honku
K'ork'ite YeTerara Senselet
Dilla
Bore
Mejo
Shela
Wenago
Suta
Bule

5

21

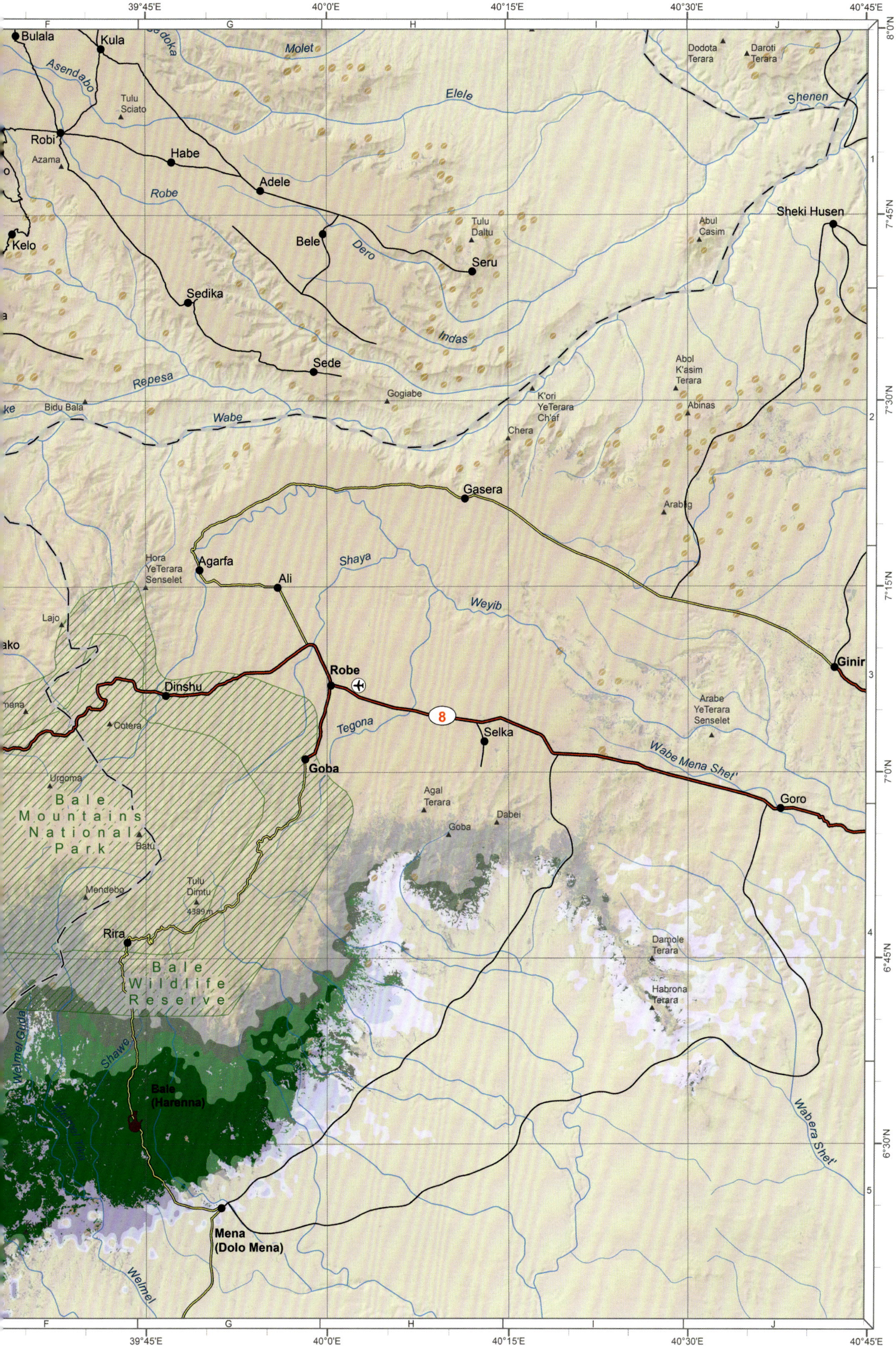

39°45'E
40°0'E
40°15'E
40°30'E
40°45'E
8°0'N
7°45'N
7°30'N
7°15'N
7°0'N
6°45'N
6°30'N
F
G
H
I
J
1
2
3
4
5
Bulala
Kula
Asendabo
Molet
Elele
Dodota Terara
Daroti Terara
Shenen
Tulu Sciato
Robi
Azama
Habe
Adele
Robe
Kelo
Bele
Dero
Tulu Daltu
Seru
Sheki Husen
Abul Casim
Sedika
Indas
Sede
Repesa
Gogiabe
Abol K'asim Terara
K'ori YeTerara Ch'af
Abinas
Bidu Bala
Wabe
Chera
Gasera
Arabig
Hora YeTerara Senselet
Agarfa
Ali
Shaya
Weyib
Lajo
Robe
Dinshu
Ginir
Cotera
Tegona
8
Selka
Arabe YeTerara Senselet
Wabe Mena Shet'
Goba
Urgoma
Goro
Bale Mountains National Park
Batu
Agal Terara
Dabei
Goba
Mendebo
Tulu Dimtu
4389m
Rira
Damole Terara
Bale Wildlife Reserve
Habrona Terara
Welmel Guda
Shawe
Bale (Harenna)
Wabera Shet'
Mena (Dolo Mena)
Welmel

14

6

22

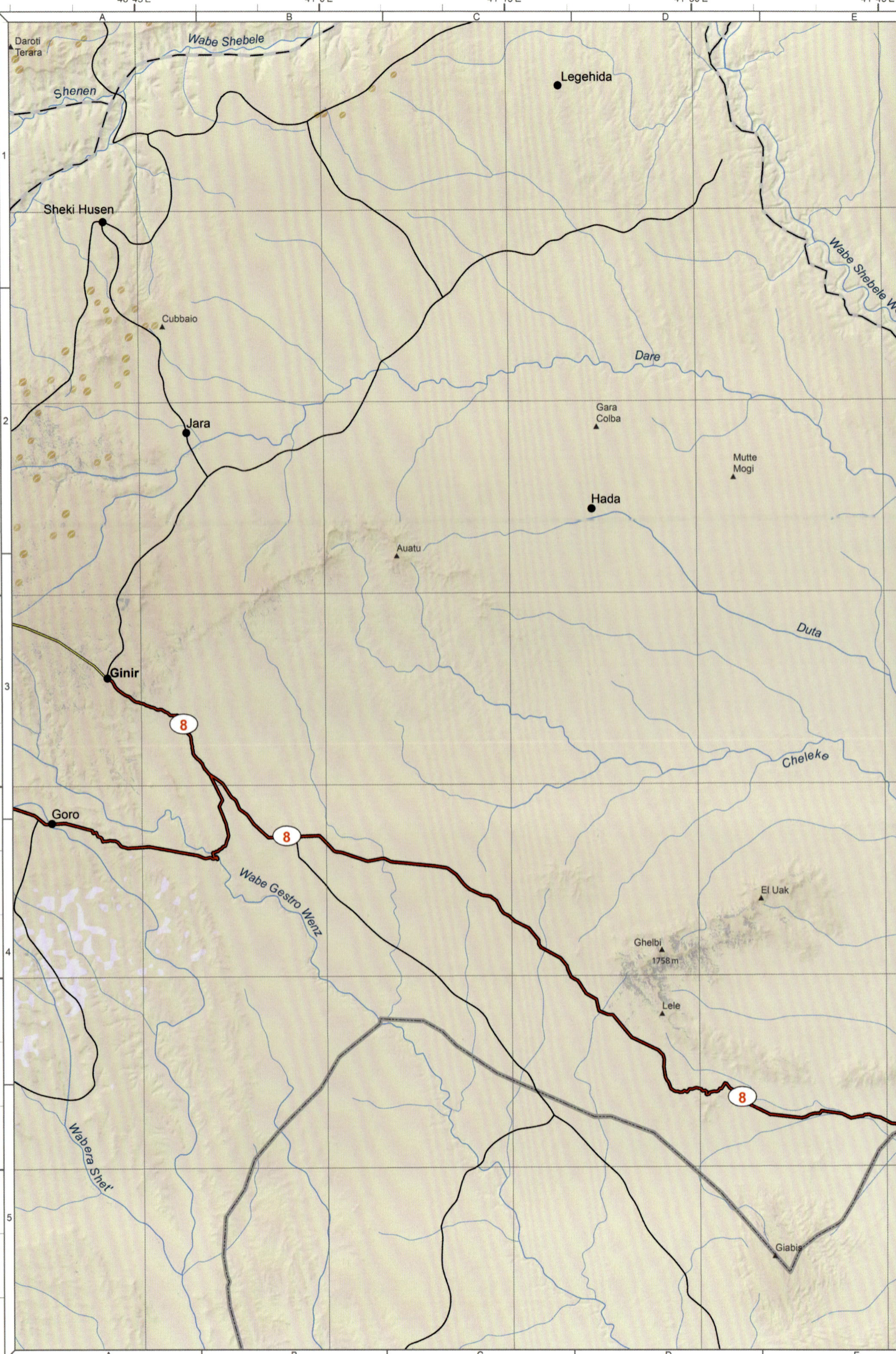

40°45'E
41°0'E
41°15'E
41°30'E
41°45'E
A
B
C
D
E
8°0'N
7°45'N
7°30'N
7°15'N
7°0'N
6°45'N
6°30'N
1
2
3
4
5
Daroti Terara
Wabe Shebele
Shenen
Legehida
Sheki Husen
Wabe Shebele Wenz
Cubbaio
Dare
Gara Colba
Jara
Mutte Mogi
Hada
Auatu
Duta
Ginir
8
Cheleke
Goro
Wabe Gestro Wenz
El Uak
Ghelbi
1758 m
Lele
Wabera Shet'
Giabis

13

7

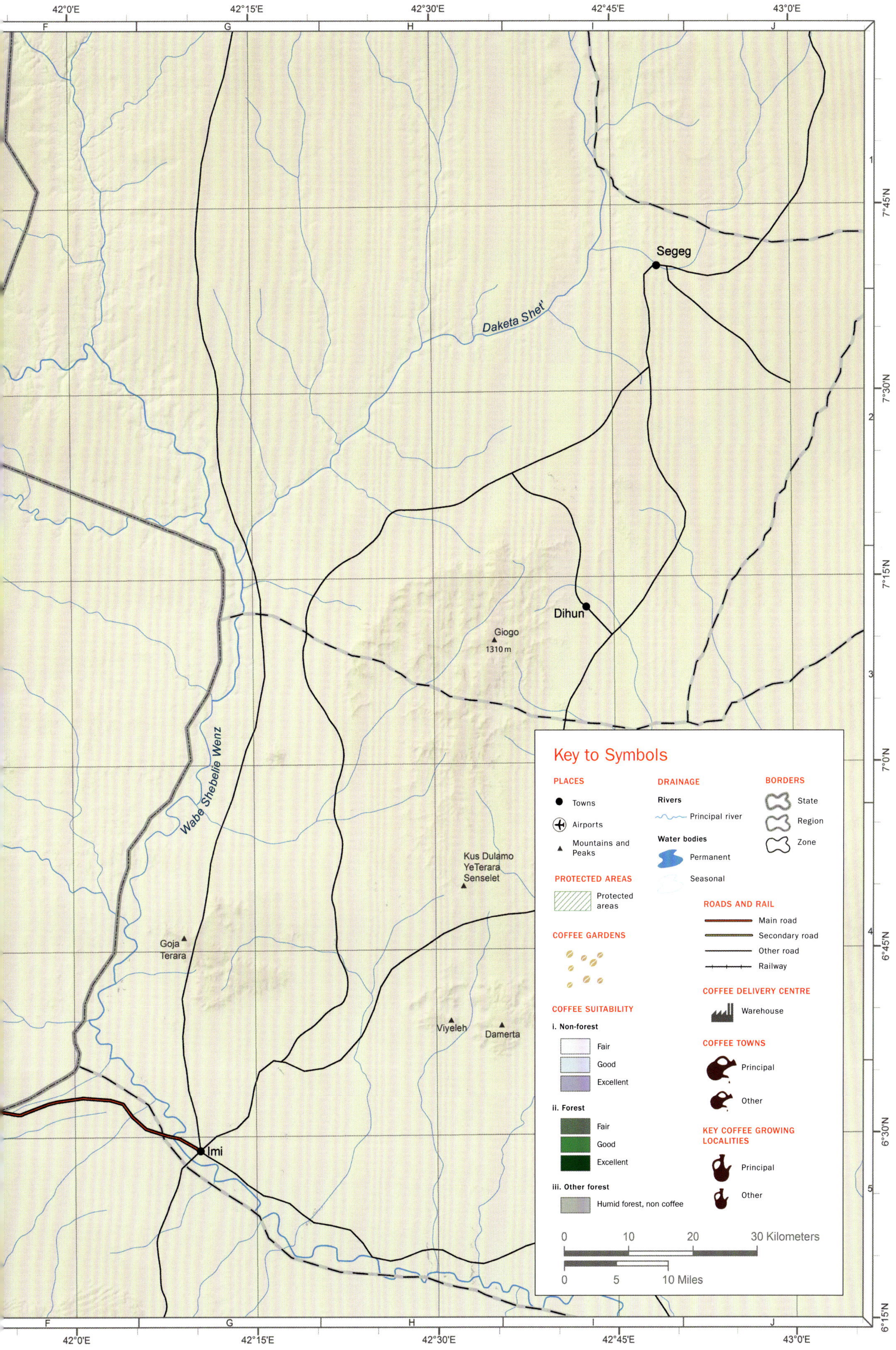
42°0'E
42°15'E
42°30'E
42°45'E
43°0'E
F
G
H
I
J
1
2
3
4
5
7°45'N
7°30'N
7°15'N
7°0'N
6°45'N
6°30'N
6°15'N
Segeg
Daketa Shet'
Dihun
Giogo
1310 m
Wabe Shebelle Wenz
Kus Dulamo
YeTerara
Senselet
Goja
Terara
Viyeleh
Damerta
Imi
Key to Symbols
PLACES
Towns
Airports
Mountains and Peaks
DRAINAGE
Rivers
Principal river
Water bodies
Permanent
Seasonal
BORDERS
State
Region
Zone
PROTECTED AREAS
Protected areas
ROADS AND RAIL
Main road
Secondary road
Other road
Railway
COFFEE GARDENS
COFFEE DELIVERY CENTRE
Warehouse
COFFEE SUITABILITY
i. Non-forest
Fair
Good
Excellent
ii. Forest
Fair
Good
Excellent
iii. Other forest
Humid forest, non coffee
COFFEE TOWNS
Principal
Other
KEY COFFEE GROWING LOCALITIES
Principal
Other
0
10
20
30 Kilometers
0
5
10 Miles

24

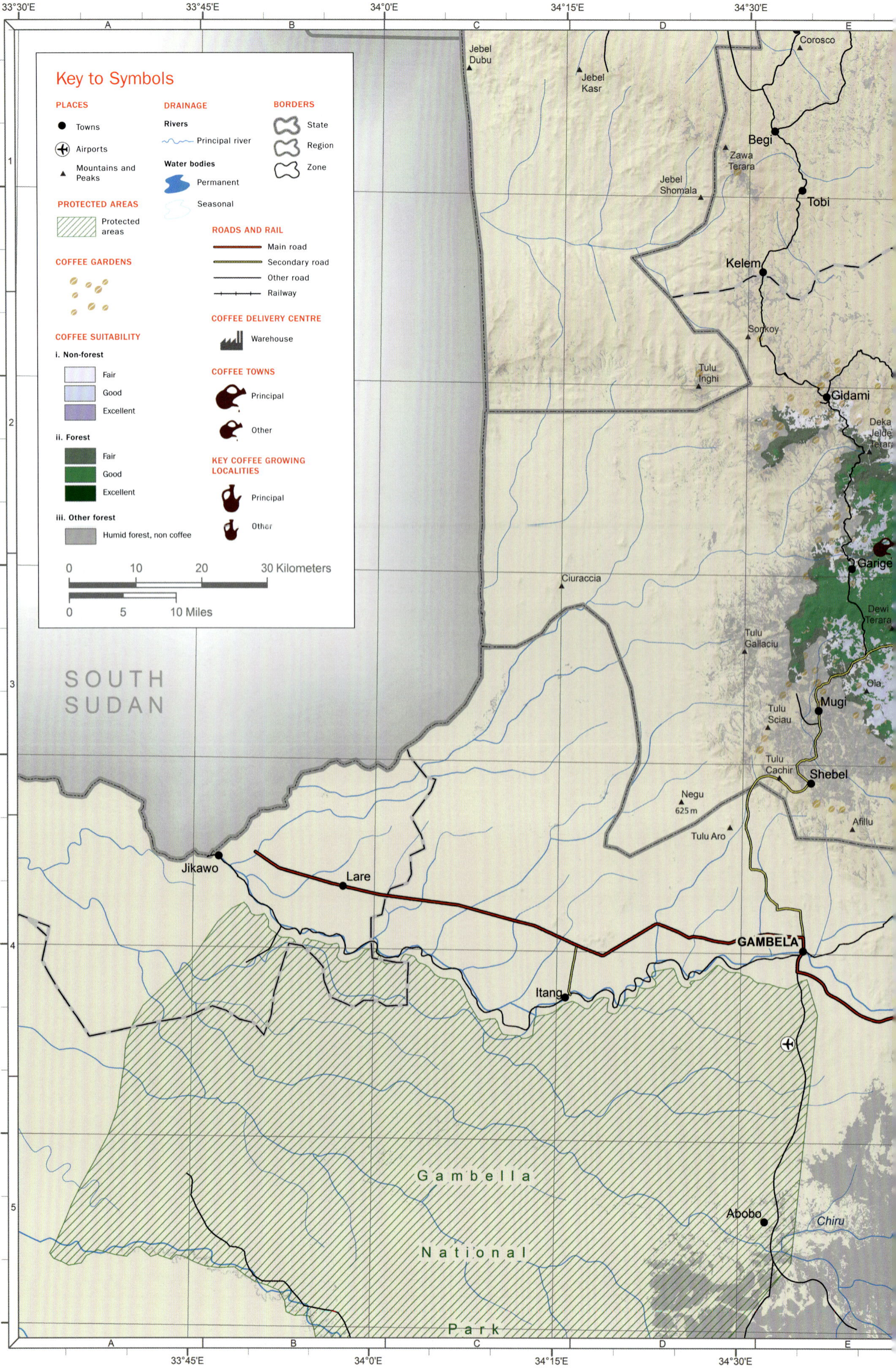
Key to Symbols
PLACES
Towns
Airports
Mountains and Peaks
PROTECTED AREAS
Protected areas
COFFEE GARDENS
COFFEE SUITABILITY
i. Non-forest
Fair
Good
Excellent
ii. Forest
Fair
Good
Excellent
iii. Other forest
Humid forest, non coffee
DRAINAGE
Rivers
Principal river
Water bodies
Permanent
Seasonal
ROADS AND RAIL
Main road
Secondary road
Other road
Railway
COFFEE DELIVERY CENTRE
Warehouse
COFFEE TOWNS
Principal
Other
KEY COFFEE GROWING LOCALITIES
Principal
Other
BORDERS
State
Region
Zone
0 10 20 30 Kilometers
0 5 10 Miles
33°30'E
33°45'E
34°0'E
34°15'E
34°30'E
9°15'N
9°0'N
8°45'N
8°30'N
8°15'N
8°0'N
7°45'N
A
B
C
D
E
1
2
3
4
5
SOUTH SUDAN
Jebel Dubu
Jebel Kasr
Corosco
Begi
Zawa Terara
Jebel Shomala
Tobi
Kelem
Sorkoy
Tulu Inghi
Gidami
Deka Jelde Terara
Garige
Ciuraccia
Dewi Terara
Tulu Galliciu
Ola
Tulu Sciau
Mugi
Tulu Cachir
Shebel
Negu
625 m
Tulu Aro
Afillu
Jikawo
Lare
GAMBELA
Itang
Abobo
Chiru
Gambella
National
Park

8

25

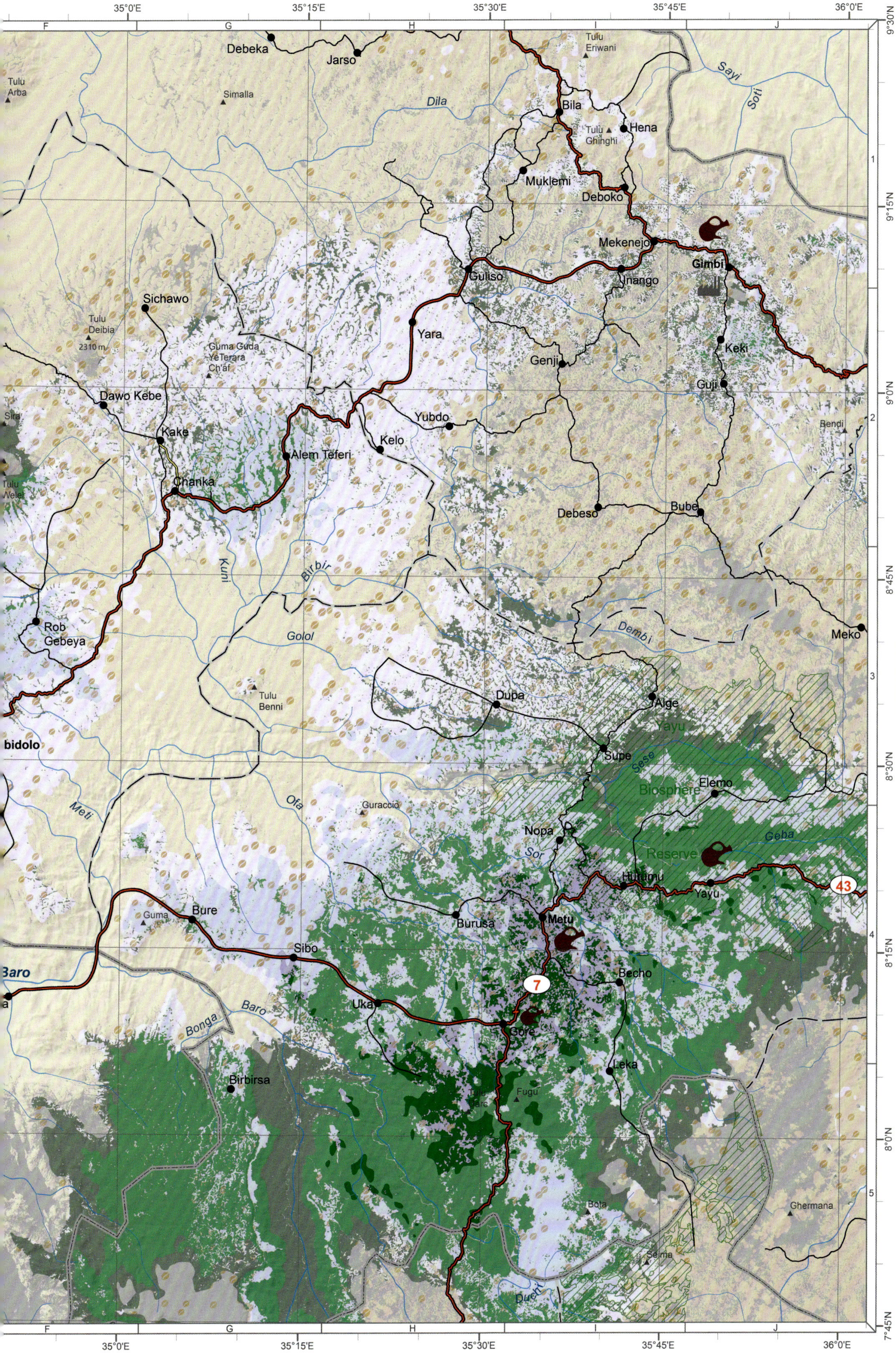
35°0'E
35°15'E
35°30'E
35°45'E
36°0'E
9°30'N
9°15'N
9°0'N
8°45'N
8°30'N
8°15'N
8°0'N
7°45'N
F
G
H
J
1
2
3
4
5
Debeka
Jarso
Tulu Eriwani
Tulu Arba
Simalla
Dila
Bila
Hena
Tulu Ghinghi
Sayi
Soti
Muklemi
Deboko
Mekenejo
Gimbi
Guliso
Inango
Sichawo
Tulu Deibia
2310 m
Guma Guda YeTerara Ch'af
Yara
Genji
Keki
Guji
Dawo Kebe
Yubdo
Kake
Alem Teferi
Kelo
Bendi
Sira
Tulu Welel
Chanka
Debeso
Bube
Kuni
Birbir
Rob Gebeya
Golol
Dembi
Meko
Tulu Benni
Dupa
Alge
bidolo
Yayu
Supe
Sese
Elemo
Biosphere
Meti
Ofa
Guraccio
Nopa
Sor
Geba
Reserve
Hurumu
Yayu
43
Guma
Bure
Burusa
Metu
Baro
Sibo
Becho
7
Baro
Bonga
Uka
Gore
Birbirsa
Leka
Fugu
Bota
Ghermana
Selma
Duchi

18

9

26

36°0'E
36°15'E
36°30'E
36°45'E
37°0'E
9°15'N
9°0'N
8°45'N
8°30'N
8°15'N
8°0'N
7°45'N
A
B
C
D
E
1
2
3
4
5
Terara
Delgagala
Shenkori Terara
SumbiShet'
Soti
Uke Wenz
Horo Shet'
Dok'onu Shet'
Hambule Terara
Gor
Wata Terara
Aua Shet'
Loko Wenz
Aieltu Wenz
Dire YeTerara Senselet
Gushis YeTerara Ch'af
Bila
Laca
Daleti Terara
Kedo YeTerara Ch'af
Gotto Muri
Marecce
Golo
Tankara
Komto Terara
Ch'alchis Terara
Rob Gebeya
Abalo
Adami
Nekemte
Mote Terara
Uakorma
Ano
Hamus Gebeya
Jogir
Arjo
Digga
Gonjo
Kolobo
Gute
Sire
Ch'ok'orsa Shet'
Fechase
Bendi
Kewisa
Jari
Dapo Gumbi
Bandiro
Sifa
Uera
Soddu
Getema
Bilo
43
Mote
Socso
Arjo
Didesa Wenz
Dabo Gumbi
Tulema
Rogghe
Arjo
Nunu
Wama Adere
Kone
Meko
Guto
Cabba
Wama
Meko
Madugga
Dega
Koye
Urgesa
Lafto
Bedele
Badda Ummo
Kemise
Aba Bor
Koma
Misingo
Geba
Dabena
Belida
Darimu
Kumbabe
Abdela
43
Bunno
Gechi
Auangero
Bekok
Tinishu Gibe
Atnago
Boia
Mile
Bido
Yanfa
Hesa
Uace
Codda
Seka
Yembero
Babo
Saie
Mulo
Iseya
Sineso
Dembi
Gent
Ucha
Setema
Ageyo
Toba
Ambuye
Onja
Sonte
Dogaia
Choche
Limu
Ghermana 2655 m
Sigamo
Bore
Agaro
Gembe
Kemise
Beshash
Gera
Yebu

17

10

27

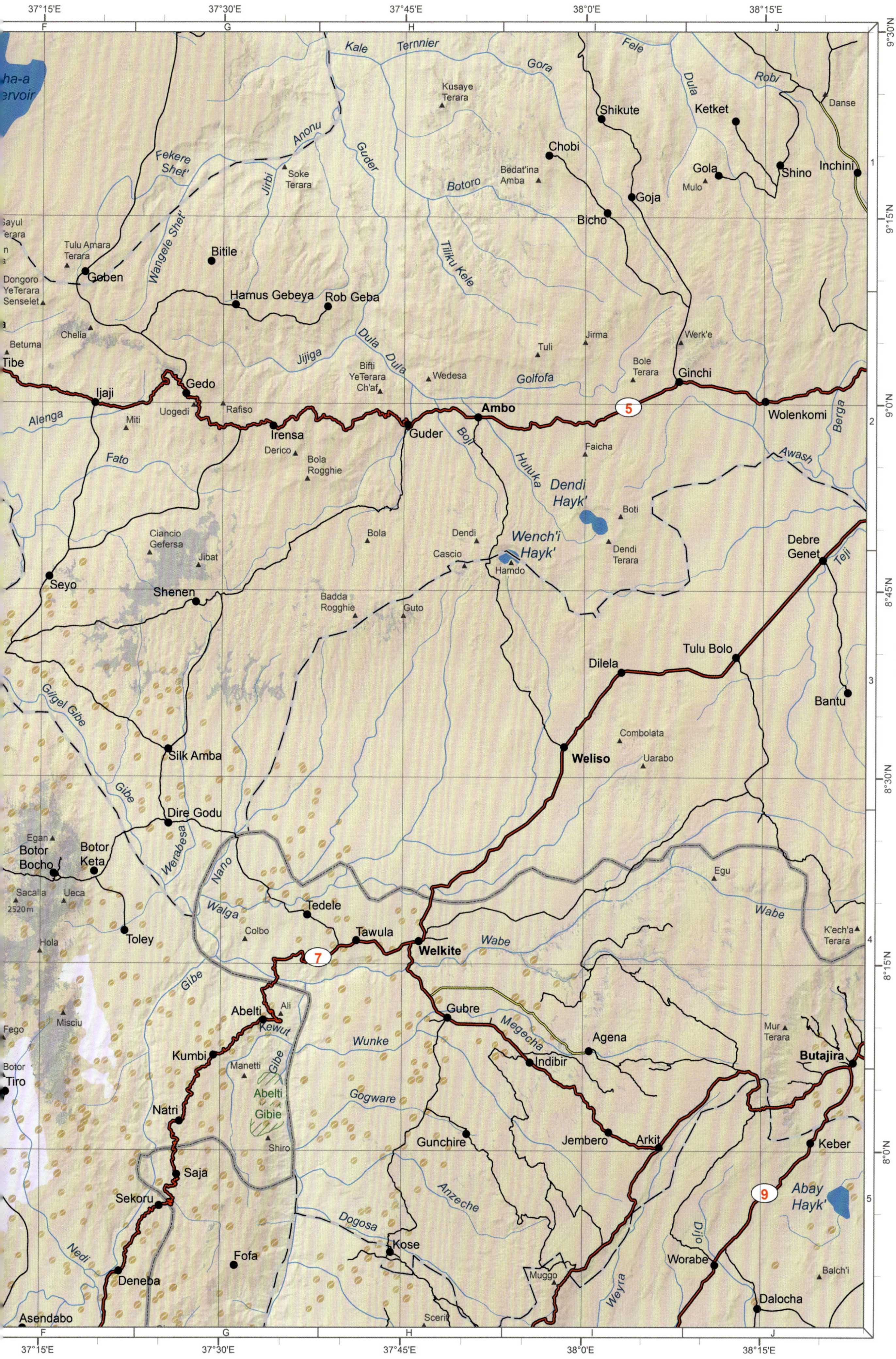
37°15'E
37°30'E
37°45'E
38°0'E
38°15'E
9°30'N
9°15'N
9°0'N
8°45'N
8°30'N
8°15'N
8°0'N
Kale
Ternnier
Gora
Fele
Dula
Robi
Danse
Kusaye Terara
Shikute
Ketket
Anonu
Guder
Fekere Shet'
Jirbi
Soke Terara
Chobi
Bedat'ina Amba
Botoro
Gola
Mulo
Shino
Inchini
Goja
Bicho
Tulu Amara Terara
Wangele Shet'
Bitile
Tiliku Kele
Goben
Dongoro YeTerara Senselet
Hamus Gebeya
Rob Geba
Chelia
Betuma
Jirma
Werk'e
Tibe
Jijiga
Dula
Tuli
Bifti YeTerara Ch'af
Wedesa
Golfofa
Bole Terara
Ginchi
Gedo
Ijaji
Rafiso
Uogedi
Ambo
Wolenkomi
Alenga
Miti
Irensa
Guder
Boli
Faicha
Berga
Derico
Bola Rogghie
Fato
Huluka
Awash
Dendi Hayk'
Boti
Bola
Dendi
Ciancio Gefersa
Wench'i Hayk'
Dendi Terara
Debre Genet
Cascio
Hamdo
Jibat
Teji
Seyo
Shenen
Badda Rogghie
Guto
Tulu Bolo
Dilela
Bantu
Gilgel Gibe
Combolata
Silk Amba
Weliso
Uarabo
Gibe
Dire Godu
Egan
Botor Bocho
Botor Keta
Werabesa
Nano
Egu
Sacalla 2520m
Ueca
Walga
Tedele
Wabe
Toley
Colbo
Tawula
K'ech'a Terara
Hola
Welkite
Wabe
Gibe
Ali
Gubre
Abelti
Kewut
Megecha
Fego
Misciu
Wunke
Agena
Mur Terara
Botor
Tiro
Kumbi
Manetti
Gibe
Indibir
Butajira
Abelti Gibie
Gogware
Natri
Shiro
Gunchire
Jembero
Arkit
Keber
Saja
Sekoru
Anzeche
Abay Hayk'
Dogosa
Nedi
Kose
Dijo
Fofa
Worabe
Muggo
Weyra
Balch'i
Deneba
Dalocha
Asendabo
Sceri
5
7
9
F
G
H
I
J
1
2
3
4
5

20

11

28

19

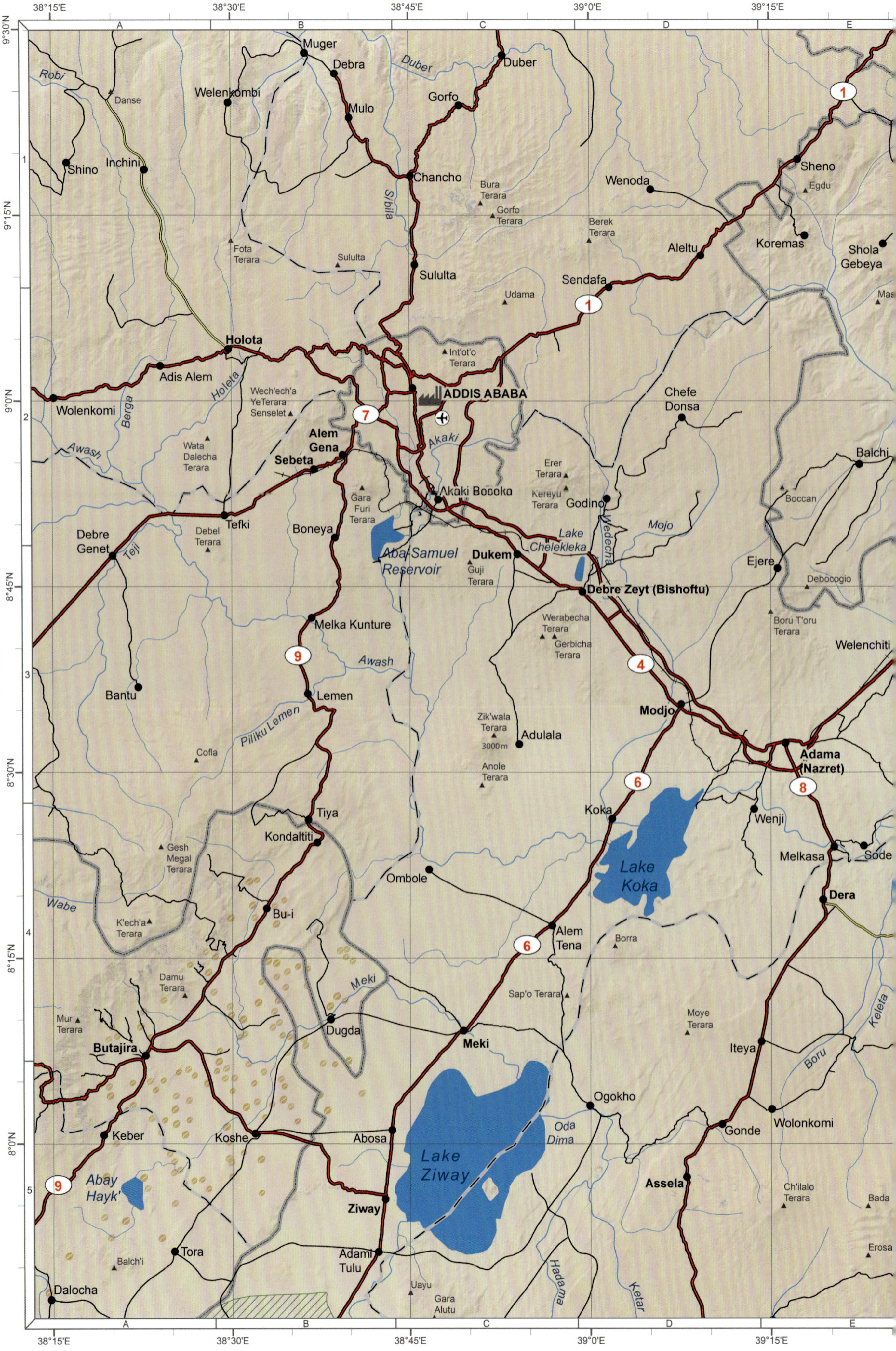
38°15'E
38°30'E
38°45'E
39°0'E
39°15'E
9°30'N
9°15'N
9°0'N
8°45'N
8°30'N
8°15'N
8°0'N
A
B
C
D
E
1
2
3
4
5
Robi
Danse
Shino
Inchini
Welenkombi
Muger
Debra
Mulo
Duber
Duber
Gorfo
Chancho
Sibila
Bura
Terara
Gorfo
Terara
Fota
Terara
Sululta
Sululta
Udama
Berek
Terara
Wenoda
Sendafa
Aleltu
Sheno
Egdu
Koremas
Shola
Gebeya
Mas
Holota
Adis Alem
Wolenkomi
Holeta
Berga
Wech'ech'a
YeTerara
Senselet
Int'ot'o
Terara
ADDIS ABABA
Chefe
Donsa
Awash
Wata
Dalecha
Terara
Alem
Gena
Sebeta
Akaki
Akaki Bosoka
Erer
Terara
Kereyu
Terara
Godino
Balchi
Boccan
Tefki
Debel
Terara
Gara
Furi
Terara
Boneya
Debre
Genet
Teji
Aba Samuel
Reservoir
Dukem
Guji
Terara
Lake
Chelekleka
Wedecha
Mojo
Ejere
Debocogio
Debre Zeyt (Bishoftu)
Melka Kunture
Awash
Werabecha
Terara
Gerbicha
Terara
Boru T'oru
Terara
Welenchiti
Bantu
Lemen
Piliku Lemen
Zik'wala
Terara
3000 m
Adulala
Modjo
Adama
(Nazret)
Cofla
Anole
Terara
Koka
Wenji
Tiya
Kondaltiti
Gesh
Megal
Terara
Lake
Koka
Melkasa
Sode
Ombole
Dera
Wabe
K'ech'a
Terara
Bu-i
Alem
Tena
Borra
Damu
Terara
Meki
Sap'o Terara
Moye
Terara
Keleta
Mur
Terara
Dugda
Meki
Butajira
Iteya
Boru
Ogokho
Keber
Koshe
Abosa
Oda
Dima
Lake
Ziway
Gonde
Wolonkomi
Abay
Hayk'
Assela
Ch'ilalo
Terara
Bada
Ziway
Tora
Adami
Tulu
Erosa
Balch'i
Hadama
Ketar
Uayu
Gara
Alutu
Dalocha
1
4
6
6
7
8
9
9
9

12

29

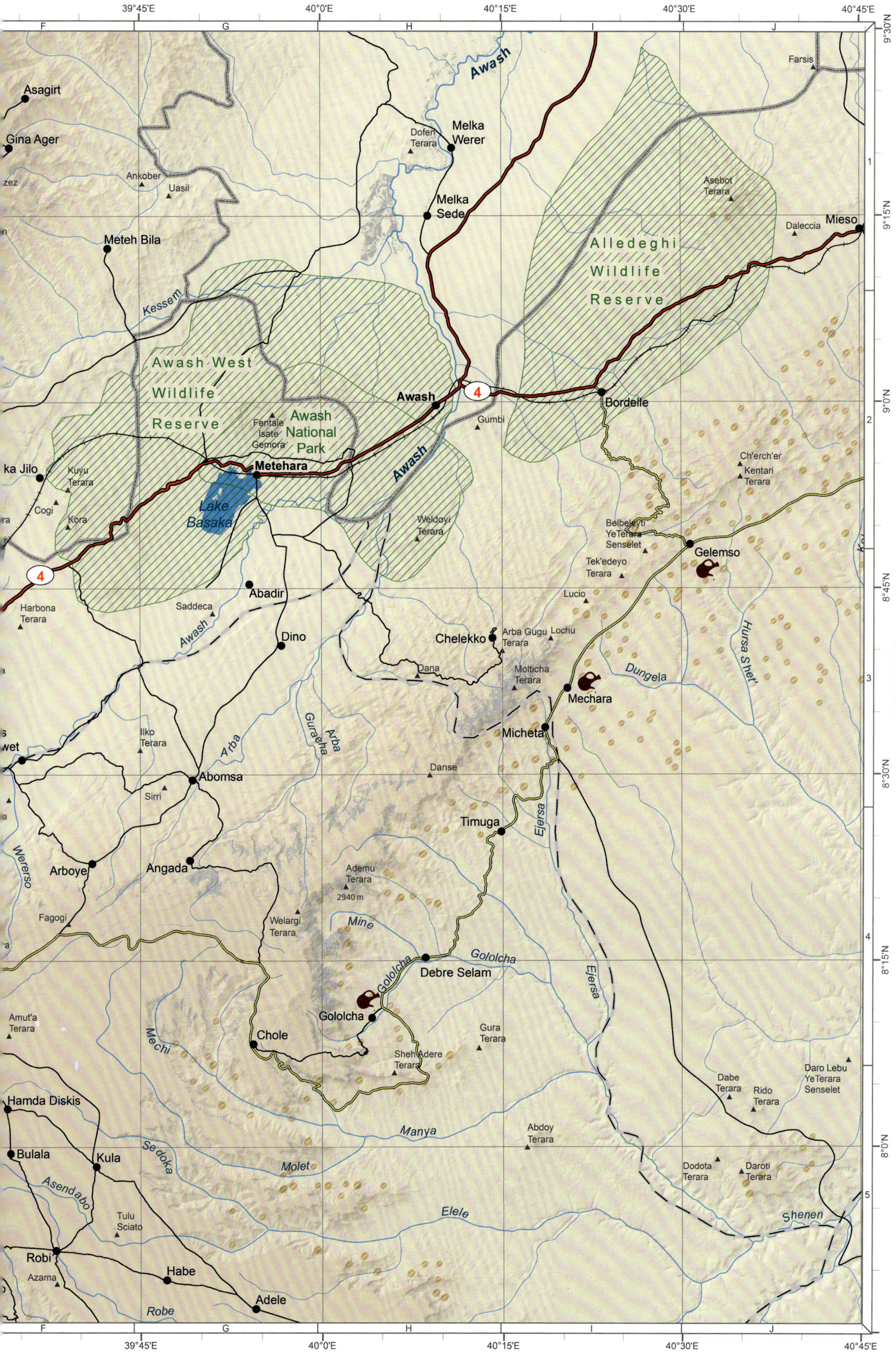
39°45'E
40°0'E
40°15'E
40°30'E
40°45'E
9°30'N
9°15'N
9°0'N
8°45'N
8°30'N
8°15'N
8°0'N
F
G
H
I
J
1
2
3
4
5
Asagirt
Gina Ager
Ankober
Uasil
Meteh Bila
Kessem
Awash
Dofen Terara
Melka Werer
Melka Sede
Farsis
Asebot Terara
Mieso
Daleccia
Alledeghi Wildlife Reserve
Awash West Wildlife Reserve
Awash National Park
Fentale Isate Gemora
Awash
4
Bordelle
Gumbi
Metehara
Lake Basaka
ka Jilo
Kuyu Terara
Cogi
Kora
Harbona Terara
Ch'erch'er
Kentari Terara
Weldoyi Terara
Belbeleyti YeTerara Senselet
Gelemso
Tek'edeyo Terara
Lucio
Abadir
Saddeca
Dino
Chelekko
Arba Gugu Terara
Lochu
Dana
Molticha Terara
Mechara
Dungela
Hursa Shet'
Micheta
Arba Guraeha
Arba
Ilko Terara
Abomsa
Sirri
Danse
Timuga
Ejersa
Wererso
Arboye
Angada
Ademu Terara
2940 m
Fagogi
Welargi Terara
Mine
Gololcha
Debre Selam
Gololcha
Amut'a Terara
Mechi
Chole
Sheh Adere Terara
Gura Terara
Dabe Terara
Rido Terara
Daro Lebu YeTerara Senselet
Hamda Diskis
Bulala
Kula
Sedoka
Molet
Manya
Abdoy Terara
Dodota Terara
Daroti Terara
Asendabo
Tulu Sciato
Elele
Shenen
Robi
Azama
Habe
Adele
Robe

22

13

30
40°45'E
41°0'E
41°15'E
41°30'E
41°45'E
9°30'N
9°15'N
9°0'N
8°45'N
8°30'N
8°15'N
8°0'N
7°45'N
Afdem
Farsis
Dalo Terara
Dembi Terara
Gulube Terara
Ahmar Mountains
Kulubi
Chelenko
Kobo
Djemi
Water
Deder
GebaShet'
Daleccia
Mieso
Hirna
Harawacha
Ala
Gara Mu
Hunde Lafto
4
Gara K'ufa
Dunk'uk'u Terara
Adem Gadi Burk'a Terara
Asbe Teferi
Mesela
Bedeno
Arba Reketi
Unde
Jelo Terara
Ch'erch'er Terara
2935 m
Mutar YeTerara Senselet
Abenyo Terara
Tortora YeTerara Senselet
Deneba Shet'
Ch'erch'er Kentari Terara
Bedesa
Kola Shet'
Sebro Terara
Umecho Shet'
Dimbiba YeTerara Senselet
Dugo Shet'
Majo Weldya
Goroga
Hursa Shet'
21
Sak'eta Shet'
Ramis Shet'
DungelaShet'
Furdo Terara
Mojo Shet'
Daro Lebu YeTerara Senselet
Rido Terara
Gedera Terara
Bilik'a Terara
Sacche
Daroti Terara
Wabe Shebele
Legehida
Shenen
14

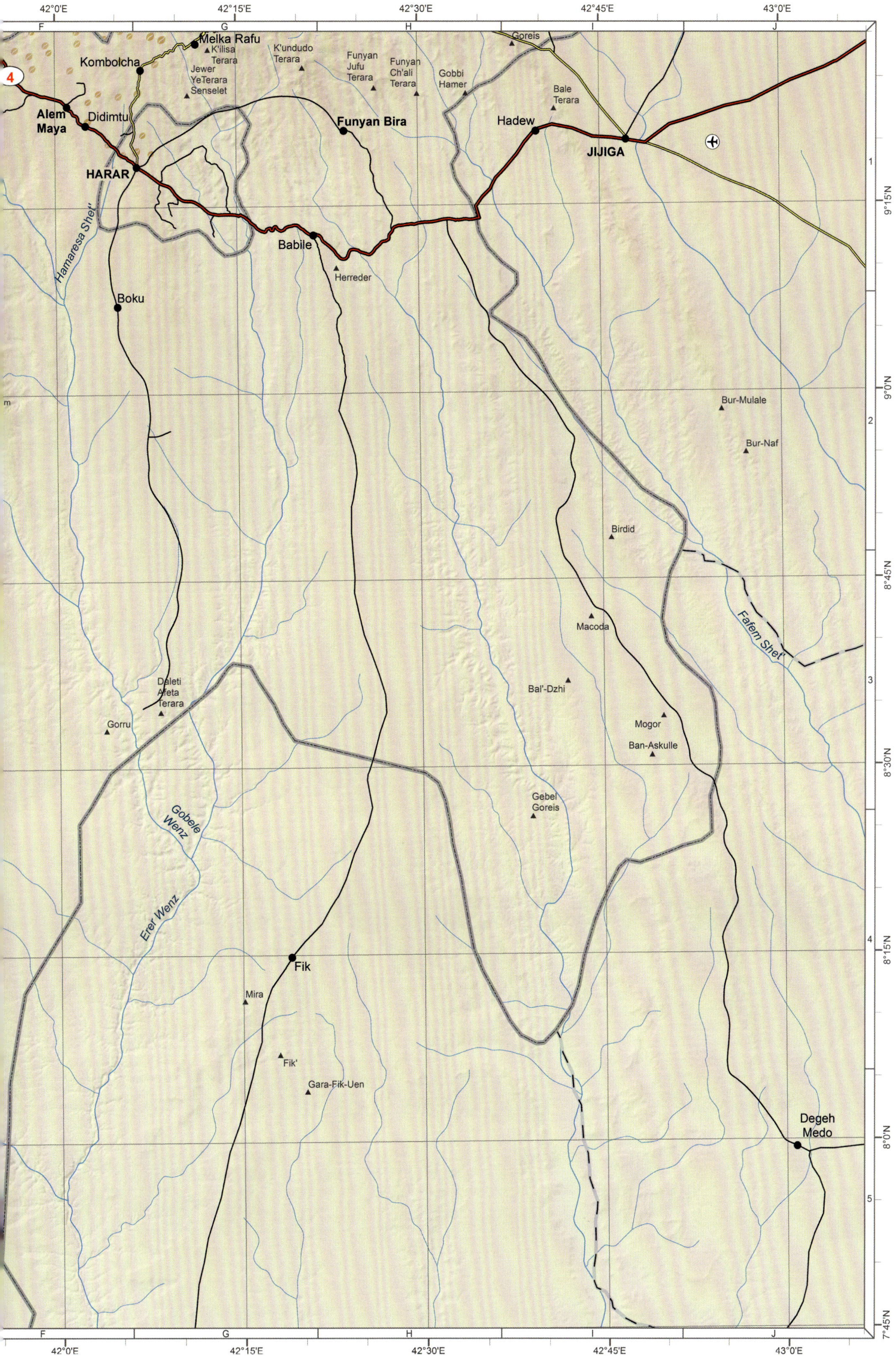
42°0'E
42°15'E
42°30'E
42°45'E
43°0'E
F
G
H
I
J
Melka Rafu
K'ilisa Terara
K'unduda Terara
Funyan Jufu Terara
Funyan Ch'ali Terara
Gobbi Hamer
Goreis
Kombolcha
Jewer YeTerara Senselet
4
Alem Maya
Didimtu
Funyan Bira
Bale Terara
Hadew
JIJIGA
HARAR
Hamaresa Shet'
Babile
Herreder
Boku
Bur-Mulale
Bur-Naf
Birdid
Macoda
Fafem Shet'
Bal'-Dzhi
Daleti Afeta Terara
Gorru
Mogor
Ban-Askulle
Gebel Goreis
Gobele Wenz
Erer Wenz
Fik
Mira
Fik'
Gara-Fik-Uen
Degeh Medo
1
2
3
4
5
9°15'N
9°0'N
8°45'N
8°30'N
8°15'N
8°0'N
7°45'N

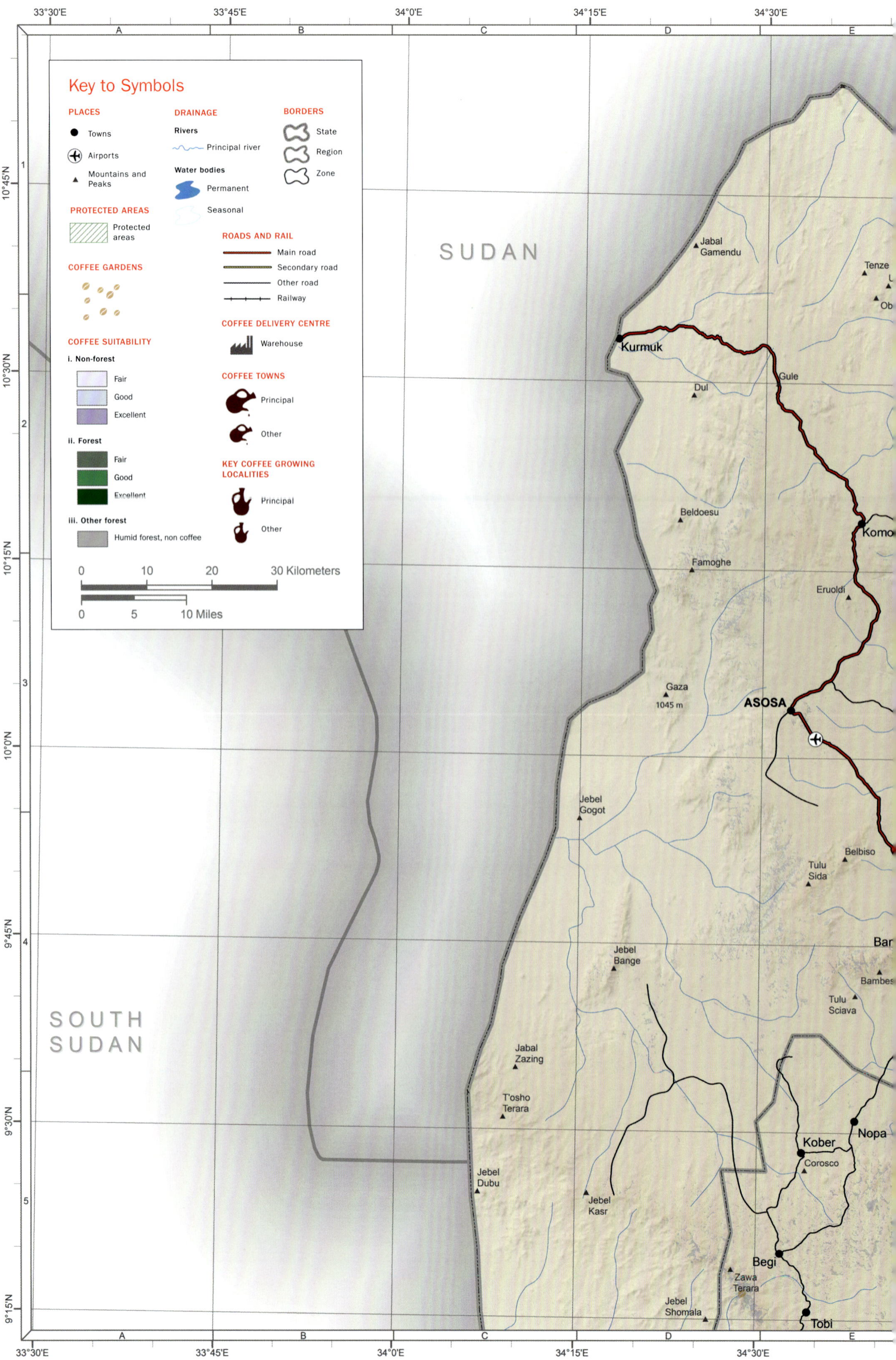

Key to Symbols
PLACES
Towns
Airports
Mountains and Peaks
DRAINAGE
Rivers
Principal river
Water bodies
Permanent
Seasonal
BORDERS
State
Region
Zone
PROTECTED AREAS
Protected areas
ROADS AND RAIL
Main road
Secondary road
Other road
Railway
COFFEE GARDENS
COFFEE DELIVERY CENTRE
Warehouse
COFFEE SUITABILITY
i. Non-forest
Fair
Good
Excellent
COFFEE TOWNS
Principal
Other
ii. Forest
Fair
Good
Excellent
KEY COFFEE GROWING LOCALITIES
Principal
Other
iii. Other forest
Humid forest, non coffee
0 10 20 30 Kilometers
0 5 10 Miles
SUDAN
SOUTH SUDAN
Jabal Gamendu
Tenze
Kurmuk
Gule
Dul
Beldoesu
Famoghe
Eruoldi
Gaza
1045 m
ASOSA
Jebel Gogot
Belbiso
Tulu Sida
Jebel Bange
Bambes
Tulu Sciava
Jabal Zazing
T'osho Terara
Jebel Dubu
Jebel Kasr
Kober
Corosco
Nopa
Begi
Zawa Terara
Jebel Shomala
Tobi
33°30'E
33°45'E
34°0'E
34°15'E
34°30'E
10°45'N
10°30'N
10°15'N
10°0'N
9°45'N
9°30'N
9°15'N

16

32

35°0'E
35°15'E
35°30'E
35°45'E
36°0'E
F
G
H
I
J
1
2
3
4
5
10°45'N
10°30'N
10°15'N
10°0'N
9°45'N
9°30'N
9°15'N
Fadokha
Jabal Belesagado
Beri
Sirba
El Danab
Mincio
Debre Zeyit (Wembera)
2580 m
Culan Sancai
Iagona
Fado
Godere
Guciar
Duma
Iagada
Huladura Terara
Fari
Dildessa
Ciababa
Fagaddo
Sirbanti Terara
Dabus
Corac
Cellalaca
Daleti
Ionghi
Ciono
Enemo
Sirba
Didesa
5
Benguwa
Mendi
Lein
Toba
Gara Nasa
Tulu Uandi
Jeldesa
Nejo
Tulu Eriwani
Debeka
Jarso
Sayi
Tulu Arba
Simalla
Dila
Bila
Soti
Tulu Ghinghi
Hena
Muklemi
Deboko

26

17

33

36°0'E
36°15'E
36°30'E
36°45'E
37°0'E
10°45'N
10°30'N
10°15'N
10°0'N
9°45'N
9°30'N
9°15'N
Sagado
Chagni
Gongi
Sigadi
Mirghi
Injibara
Fudi
Kesa
Taia
Gimja Bet
Tilili
Dibate
Azena
Darim
Kilaj
Achigi
Bulan
Zigin
Shindi
Gum
Bulen
May Dara Shet'
Gutr
Galesa
Gomer
Boramu
Gubbi
Korka
Blue Nile (Abeya Wenz)
Sirbanti Terara
Abutalla Ibantu
Kelo
Agemsa
Ueni
Corac
Jangir
Soghidda
Nefruit
Obe
Gidda
Surge
Lench'a Terara
Kiremu
Morna
Dire
Alaba Meda
Kokofe
Bora Terara
Yaso YeTerara Senselet
Gondor Terara
Ayana
Haro
Gelila
Dami
Gerchi Wenz
Jarc
Gomboch' Terara
Garamate
Mulu Terara
Sombo
Gibir
Uaggi
2110 m
Jeldu Terara
Gone Terara
Jeldesa
Waja Wenz
Lafto Terara
Sha
Hanger Wenz
Gello
Dale Terara
Ark'umbe Terara
Cuma
Serbe Dale Shet'
Koma Terara
Hambule Terara
Delgagala
Sumbi Shet'
Shenkori Terara
Uke Wenz
Horo Shet'
Dok'onu Shet'
Soti
Wata Terara
Aua Shet'
Loko Wenz
Dire YeTerara Senselet
Aieltu Wenz
Gushis YeTerara Ch'af
Bila
Laca
Daleti Terara

25

18

34

37°15'E
37°30'E
37°45'E
38°0'E
38°15'E
11°0'N
10°45'N
10°30'N
10°15'N
10°0'N
9°45'N
9°30'N
9°15'N
Gish Abay
Satima
Lich'ma Lechema
Keraniya
Gunde Weyn
Amoneuos
Medeg
Deri
Sede
Genet
Feres Bet
Digo Tsiyon
Mertule Maryam
Feres Mada
Achewa Bado
Tame
Uari
Celti
Carni
Ch'ok'e
Ch'ok'e Terara
Felege Birhan
Mankusa
Ingia
Finote Selam
Jiga
Anjeni
Urrata
Debre Werk
Mekera Kir
Daguai
Terara
Talo
3750 m
Zaua
Senyin
Arat Mekera Kir
Amora
Mengistiwo
Dembecha
Shelel
Rob Gebeya
Kuyi
Debet
Amanuel
Bichena
Yebokile
Yewla
Suha
Webrje
Yetmen
Debre Markos
Chemoga
Muga
Wamit
Eliyas
Lumame
Yekoza
3
Dejen
Yejube
Bogena
Filikilik
Gosha Tsiyon
Bete Nigus
Tulu Bedada
Yeda
Hose
Tulu Miki
Ch'ogo Wenz
Nedi Shet'
Finch'a-a Wenz
Kewo
Mitra
Fichil
Bechusa
Kere Dobi
Biriti
Uru
Dedu
Muger
Achane
Chulute
Jemo Lefo
Guder
Minare
Kachisi
Fincha-a
Degat'ina Amba
Kembolcha
Mekoda
Kombolcha
Kale
Temnier
Fele
Gora
Robi
Dula
cha-a servoir
Kusaye Terara
Danse
Shikute
Ketket
Anonu
Chobi
Fekere Shet'
Guder
Gola
Shino
Inchini
Jirbi
Soke Terara
Botoro
Bedat'ina Amba
Mulo
Goja
F
G
H
I
J
1
2
3
4
5

28

19

35

27

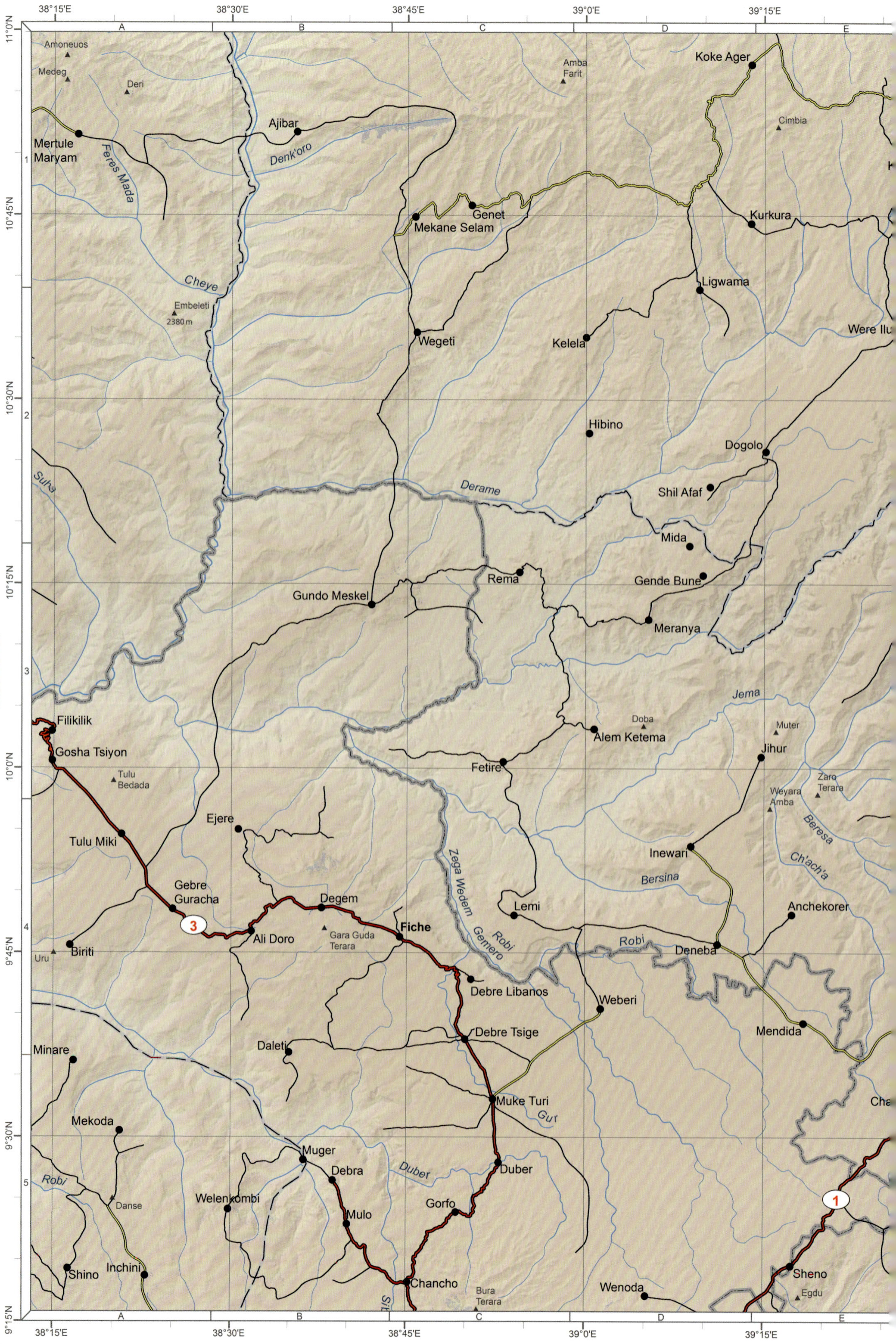
38°15'E
38°30'E
38°45'E
39°0'E
39°15'E
11°0'N
10°45'N
10°30'N
10°15'N
10°0'N
9°45'N
9°30'N
9°15'N
A
B
C
D
E
1
2
3
4
5
Amoneuos
Medeg
Deri
Amba Farit
Koke Ager
Cimbia
Ajibar
Mertule Maryam
Feres Mada
Denk'oro
Genet
Mekane Selam
Kurkura
Cheye
Ligwama
Embeleti
2380 m
Wegeti
Kelela
Were Ilu
Hibino
Dogolo
Suha
Derame
Shil Afaf
Mida
Rema
Gende Bune
Gundo Meskel
Meranya
Jema
Filikilik
Doba
Muter
Alem Ketema
Gosha Tsiyon
Jihur
Tulu Bedada
Fetire
Zaro Terara
Weyara Amba
Ejere
Beresa
Inewari
Tulu Miki
Ch'ach'a
Zega Wedem
Bersina
Gebre Guracha
Degem
Lemi
Anchekorer
3
Ali Doro
Gara Guda Terara
Fiche
Robi
Gemero
Robi
Deneba
Uru
Biriti
Debre Libanos
Weberi
Mendida
Debre Tsige
Minare
Daleti
Muke Turi
Cha
Gur
Mekoda
Muger
Duber
Debra
Duber
Robi
Welenkombi
1
Danse
Gorfo
Mulo
Shino
Inchini
Sheno
Chancho
Wenoda
Bura Terara
Egdu
Sib

20

36

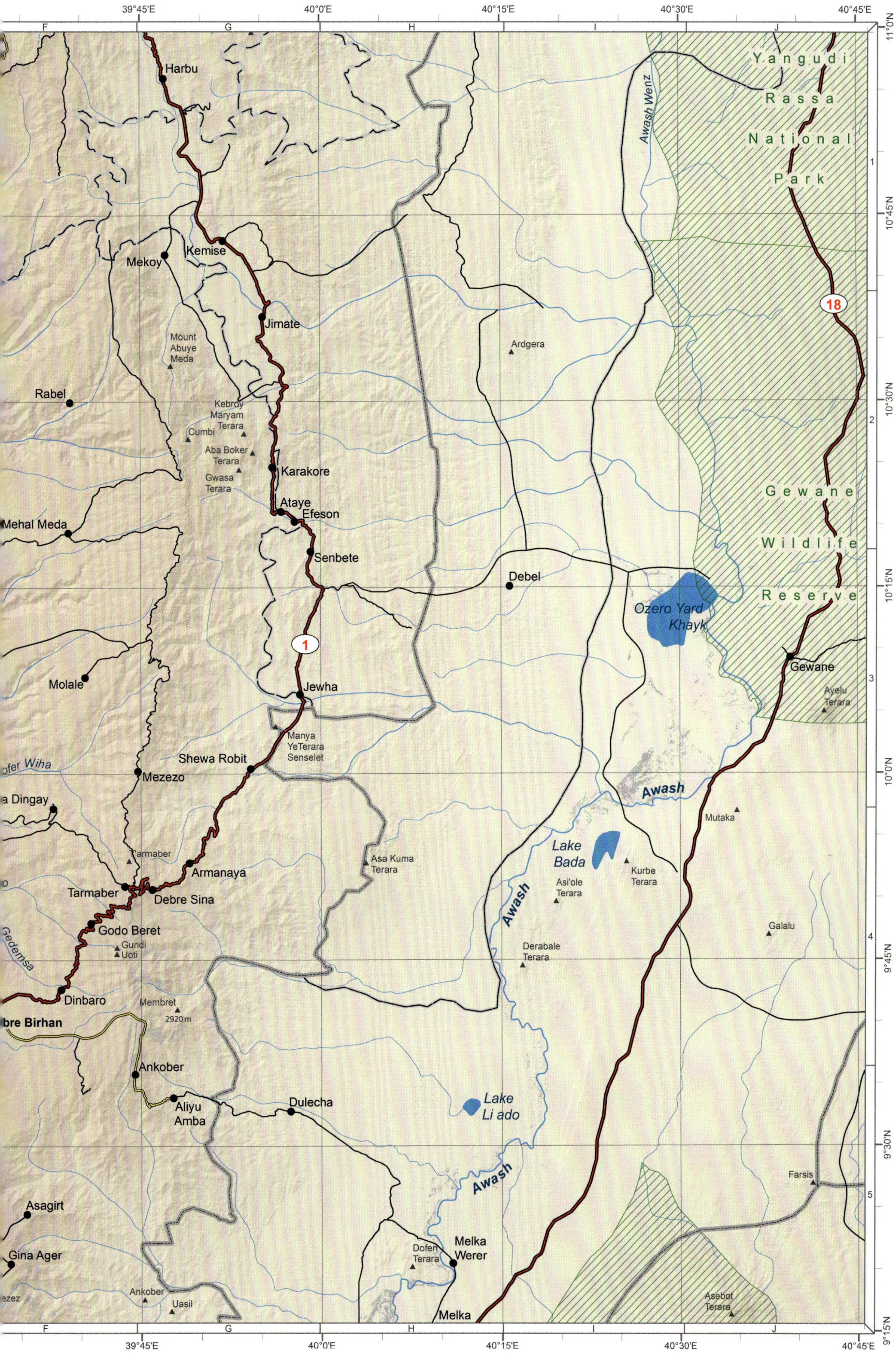
39°45'E
40°0'E
40°15'E
40°30'E
40°45'E
11°0'N
10°45'N
10°30'N
10°15'N
10°0'N
9°45'N
9°30'N
9°15'N
Yangudi Rassa National Park
Gewane Wildlife Reserve
Harbu
Kemise
Mekoy
Jimate
Mount Abuye Meda
Rabel
Kebroy Maryam Terara
Cumbi
Aba Boker Terara
Gwasa Terara
Karakore
Ataye
Efeson
Mehal Meda
Senbete
Debel
Ardgera
Awash Wenz
18
Ozero Yard Khayk
1
Molale
Jewha
Gewane
Ayelu Terara
Manya YeTerara Senselet
Shewa Robit
Mezezo
Dingay
Awash
Mutaka
Tarmaber
Armanaya
Asa Kuma Terara
Lake Bada
Kurbe Terara
Asi'ole Terara
Debre Sina
Godo Beret
Gundi Uoti
Gedemsa
Galalu
Derabale Terara
Dinbaro
Membret 2920m
bre Birhan
Ankober
Aliyu Amba
Dulecha
Lake Li ado
Farsis
Asagirt
Gina Ager
Melka Werer
Dofen Terara
Ankober
Uasil
Asebot Terara
Melka

30

21

37

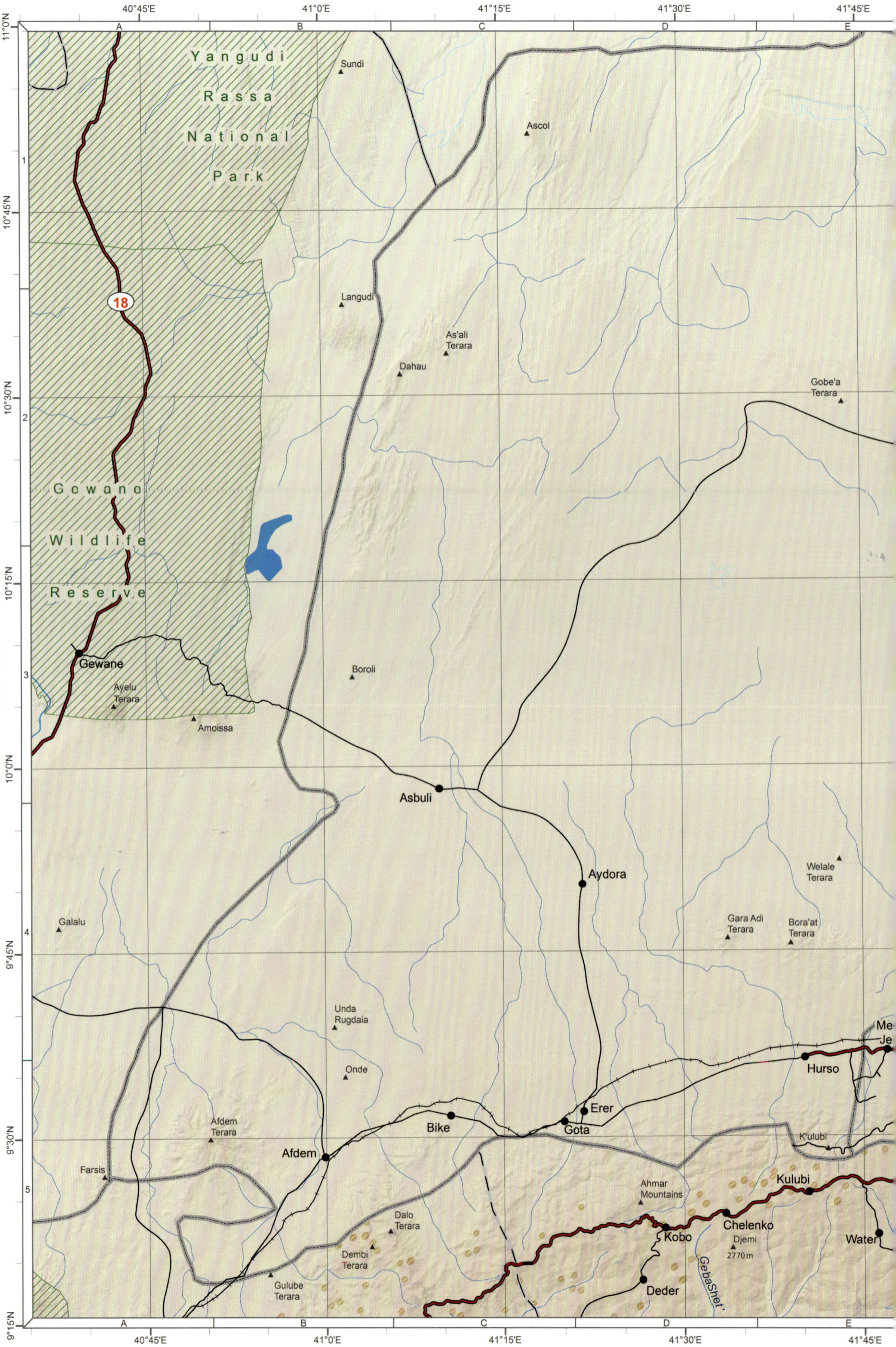
40°45'E
41°0'E
41°15'E
41°30'E
41°45'E
11°0'N
10°45'N
10°30'N
10°15'N
10°0'N
9°45'N
9°30'N
9°15'N
Yangudi Rassa National Park
Sundi
Ascol
18
Langudi
As'ali Terara
Dahau
Gobe'a Terara
Gewane Wildlife Reserve
Gewane
Ayelu Terara
Amoissa
Boroli
Asbuli
Aydora
Welale Terara
Galalu
Gara Adi Terara
Bora'at Terara
Unda Rugdaia
Onde
Hurso
Erer
Bike
Gota
Afdem Terara
Afdem
K'ulubi
Farsis
Ahmar Mountains
Kulubi
Dalo Terara
Chelenko
Kobo
Djemi
2770 m
Water
Dembi Terara
Gulube Terara
Deder
GebaShet'

29

22

42°0'E
42°15'E
42°30'E
42°45'E
43°0'E
10°45'N
10°30'N
10°15'N
10°0'N
9°45'N
9°30'N
9°15'N
SOMALIA
Ara le Koma
Dagah Lebi
Gumar
Asengedi
Asmarerle
Erene
Akaley
Burale
Godire
Arabdara
Maror
Durdur
Deryo
Arag
Mekho
Egle
Enlon-Godon
Huglo
Ayisha
Bur Degamedou Dzher
Mil'milaki-Terari
Harr
Adigala
Bur Madu
Biye Gurgur
Arome
Eles-Terara
Gogti
Biye Anod
Goba-Dzher
Doba Terara
Harawa
Arabi
Denbel
Djebi Abeyou
Salala
Amara Terara
Duleti Terara
Gara Garri
Dere Meley YeTerara Senselet
Darimi YeTerara Senselet
Kerora YeTerara Senselet
Dananley
Bulloynko
Lefe Isa
Souboul Hanfele
E DAWA
Wedaja Terara
Lalu Terara
Ink'uftu YeTerara Senselet
Dibriuein
Moccanis
Hama YeTerara Senselet
Gara Amola
Chinaksen
Were Mucha Terara
Berita Terara
Gara Goreis
Melka Rafu
K'ilisa Terara
K'undudo Terara
2950 m
Funyan Jufu Terara
Funyan Ch'ali Terara
Gobbi Hamer
Kombolcha
Jewer YeTerara Senselet
Bale Terara
Alem Maya
Didimtu
Funyan Bira
Hadew
JIJIGA
HARAR
4
23

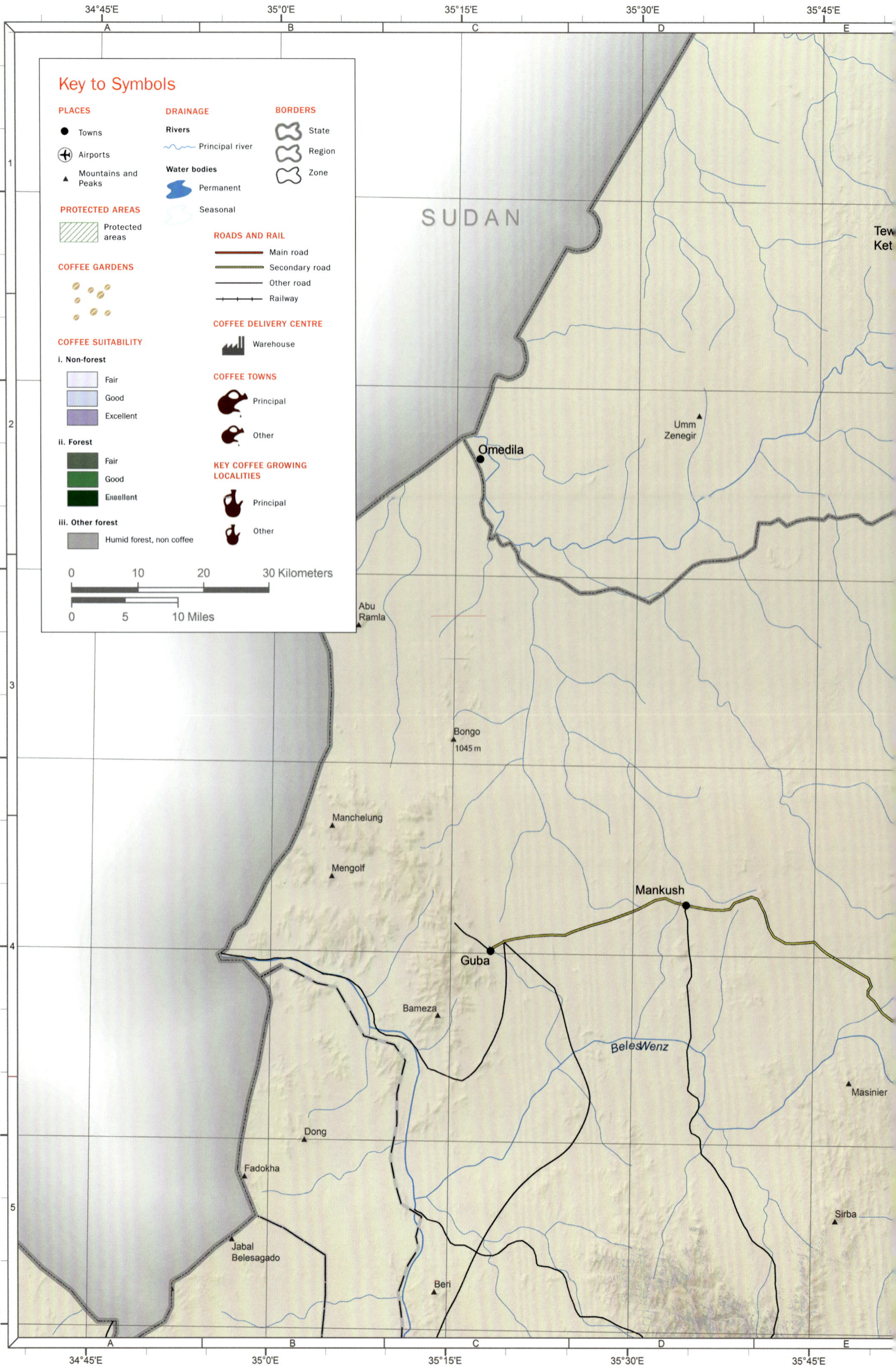

Key to Symbols
PLACES
Towns
Airports
Mountains and Peaks
PROTECTED AREAS
Protected areas
COFFEE GARDENS
COFFEE SUITABILITY
i. Non-forest
Fair
Good
Excellent
ii. Forest
Fair
Good
Excellent
iii. Other forest
Humid forest, non coffee
DRAINAGE
Rivers
Principal river
Water bodies
Permanent
Seasonal
BORDERS
State
Region
Zone
ROADS AND RAIL
Main road
Secondary road
Other road
Railway
COFFEE DELIVERY CENTRE
Warehouse
COFFEE TOWNS
Principal
Other
KEY COFFEE GROWING LOCALITIES
Principal
Other
0 10 20 30 Kilometers
0 5 10 Miles
34°45'E
35°0'E
35°15'E
35°30'E
35°45'E
A
B
C
D
E
1
2
3
4
5
12°15'N
12°0'N
11°45'N
11°30'N
11°15'N
11°0'N
10°45'N
SUDAN
Omedila
Umm Zenegir
Abu Ramla
Bongo
1045 m
Manchelung
Mengolf
Mankush
Guba
Bameza
Beles Wenz
Masinier
Dong
Fadokha
Jabal Belesagado
Beri
Sirba

25

38

36°0'E
36°15'E
36°30'E
36°45'E
37°0'E
12°30'N
12°15'N
12°0'N
11°45'N
11°30'N
11°15'N
11°0'N
10°45'N
F
G
H
I
J
1
2
3
4
5
Alcadid
Amfelau
Cuora
Galea
Chandiba
Quara
Umm
Barberi
Cura
Luza
Abracioc
Fuldana
Delgi
Gaz Gimb
Jebel Umm
Zenegia
Gilfu
Jamifala
Rahad
Lake
Tana
Amistiya
Garasgh
Bardanchi
Gandi
Dinqil Ber
Gabina
Bafa
Jagaighi
Tarara
Kunzila
Saba
Elfing
Bar
Gilgil Abay
Mosabia
Yismala
Sena
Mara
Belaya
2700 m
Merawi
Durbete
Wetet Abay
3
Mambuk
Saguma
Pawe
Dangla
Ailungiucsa
Mandura
Adis Kidane
Fogota
Assem
Sellasie
Kidamaja
Sagado
Gongi
Mirghi
Injibara
Chagni
Sigadi
Fudi
Kesa
Taia
Gimja Bet
Tilili
Agut
Dibate
Azena
36°15'E
36°30'E
36°45'E
37°0'E

34

26

37°15'E
37°30'E
37°45'E
38°0'E
12°30'N
12°15'N
12°0'N
11°45'N
11°30'N
11°15'N
11°0'N
10°45'N
Belessen
Chandiba
Kola Diba
Meksenyit
Guhala
Zoz
Asca
Chwahit Mender
Robit
Tulu Ghereccia
Chendam
Uadagi
Dumi
Uenfit
Dauc
Gorgora
Safad
Gwaliya
Delgi
Adis Zemen
Asiba
Ezub
Samo
Ibnat
Lake Tana
Yifag
Azer
Irb Shet'
Amora Gedel
Chelala
Melza
Woreta
Amed Ber
Kunzila
Debre Tabor
Hamusit Mender
Wanzaye
Gasay
Zege (Zeghie)
Anbesame
Gilgil Abay
Mize
Arb Gebeya
Mekane Eyesus
Gina
BAHIR DAR
Debra Chente
Tis Abay
Meshenti
Merawi
Mekari
Jara Gedo
Wetet Abay
Agta
Abola
Adet
Gonj
Amedamit
Lieu
3210 m
Lijambera
Mota
Quarib
Baza Shet'
Fogota
Mizan
Dabi
Gish Abay
Satima
Lich'ma Lechema
Keraniya
Gunde Weyn
Sede
Agut
Genet
Feres Bet
Digo Tsiyon
Achewa Bado
Tame
Uari
Celti
Uajza

33

40

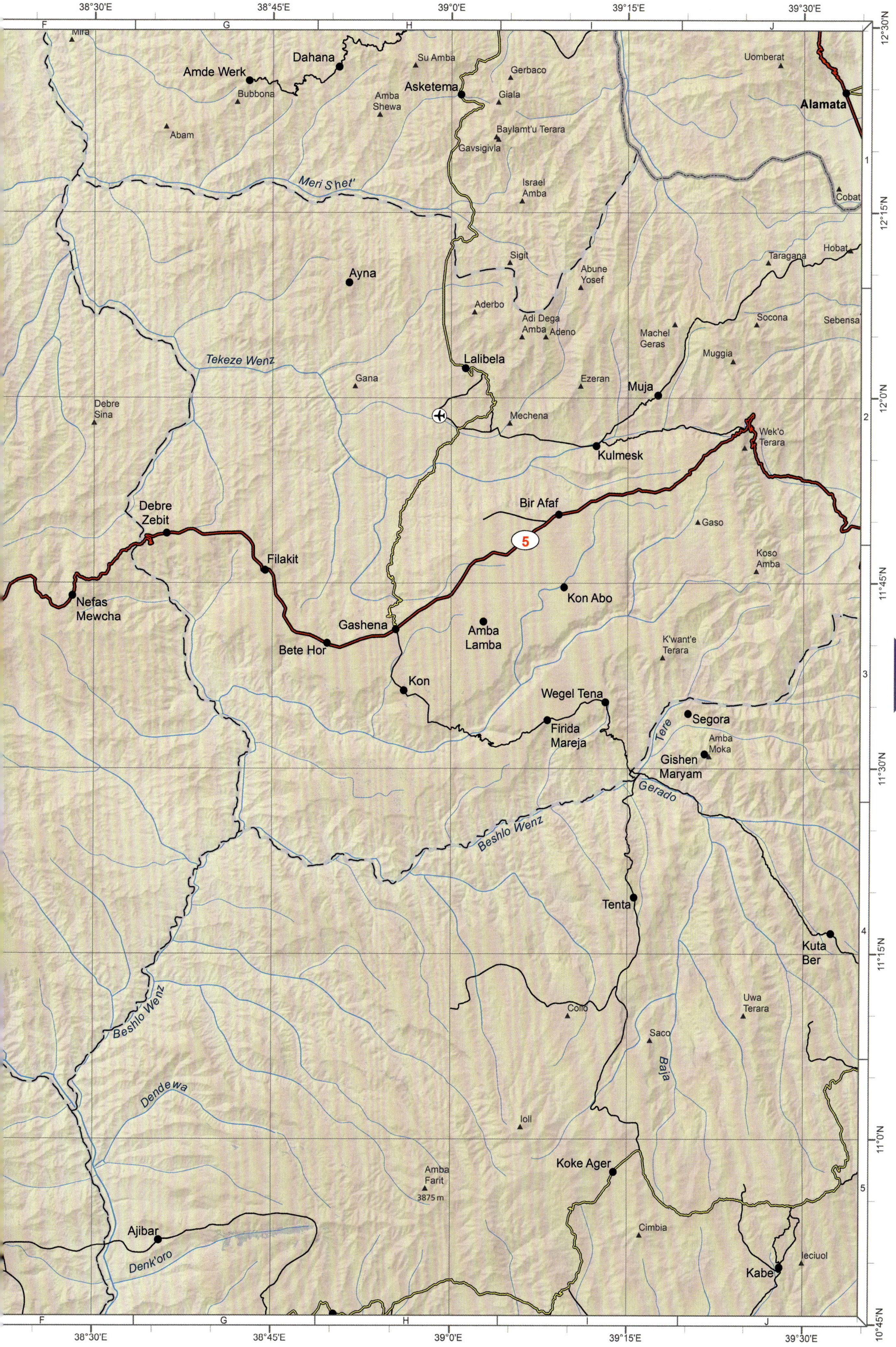
38°30'E
38°45'E
39°0'E
39°15'E
39°30'E
12°30'N
12°15'N
12°0'N
11°45'N
11°30'N
11°15'N
11°0'N
10°45'N
F
G
H
I
J
1
2
3
4
5
Mira
Dahana
Amde Werk
Bubbona
Su Amba
Asketema
Amba Shewa
Gerbaco
Giala
Uomberat
Alamata
Abam
Baylamt'u Terara
Gavsigivla
Meri S'het'
Israel Amba
Cobat
Hobat
Taragana
Sigit
Abune Yosef
Ayna
Aderbo
Adi Dega Amba
Adeno
Machel Geras
Socona
Sebensa
Muggia
Tekeze Wenz
Lalibela
Gana
Ezeran
Muja
Debre Sina
Mechena
Wek'o Terara
Kulmesk
Bir Afaf
Gaso
Debre Zebit
5
Filakit
Koso Amba
Nefas Mewcha
Kon Abo
Gashena
Amba Lamba
Bete Hor
K'want'e Terara
Kon
Wegel Tena
Segora
Firida Mareja
Tere
Amba Moka
Gishen Maryam
Gerado
Beshlo Wenz
Tenta
Kuta Ber
Beshlo Wenz
Uwa Terara
Collo
Saco
Baja
Dendewa
Ioll
Koke Ager
Amba Farit
3875 m
Cimbia
Ajibar
Denk'oro
Ieciuol
Kabe

36

28

41

35

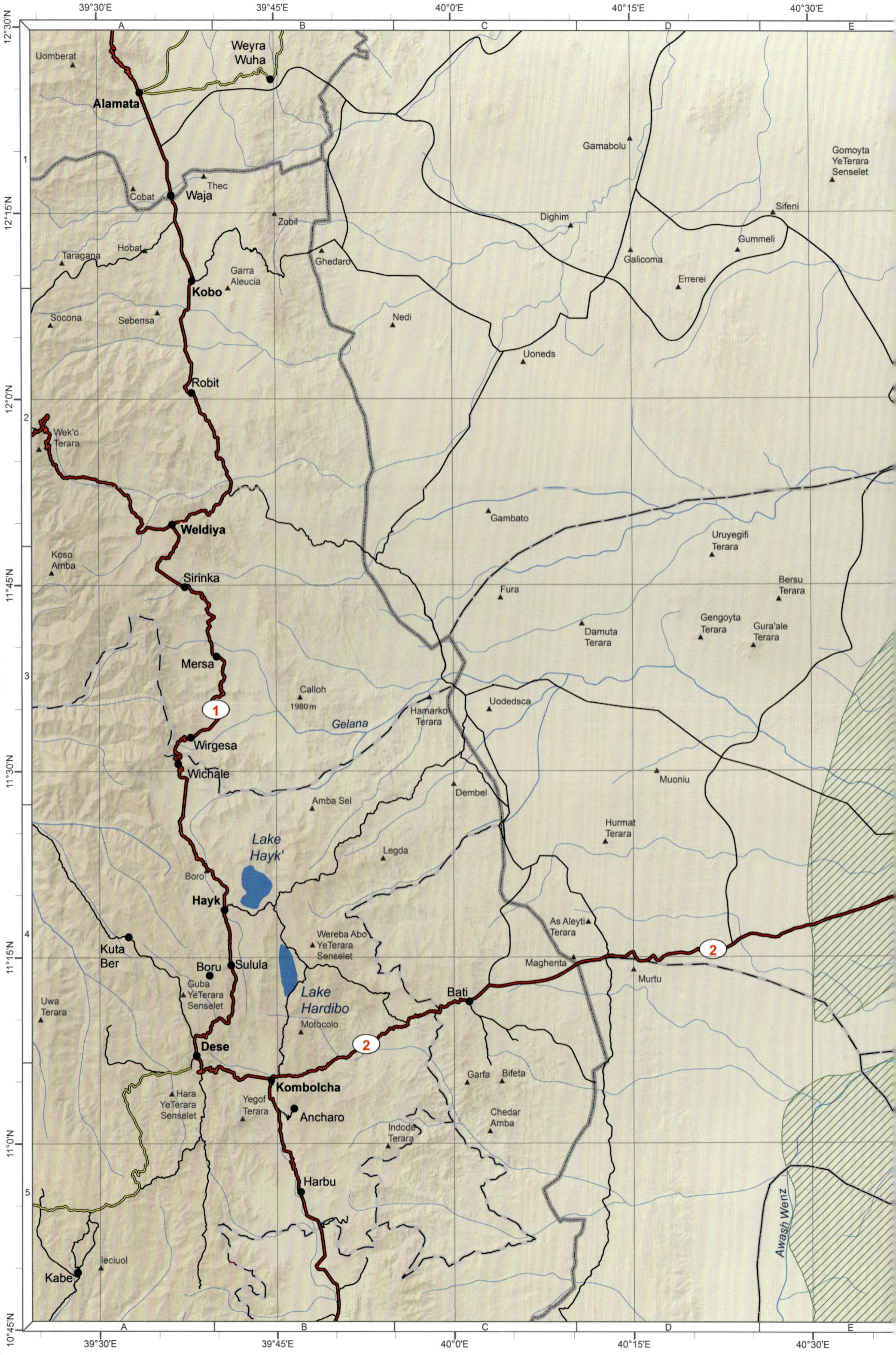

39°30'E
39°45'E
40°0'E
40°15'E
40°30'E
12°30'N
12°15'N
12°0'N
11°45'N
11°30'N
11°15'N
11°0'N
10°45'N
A
B
C
D
E
1
2
3
4
5
Weyra Wuha
Uomberat
Alamata
Gamabolu
Gomoyta YeTerara Senselet
Cobat
Thec
Waja
Zobil
Dighim
Sifeni
Gummeli
Taragana
Hobat
Ghedaro
Galicoma
Errerei
Garra Aleucia
Kobo
Socona
Sebensa
Nedi
Uoneds
Robit
Wek'o Terara
Weldiya
Gambato
Uruyegifi Terara
Koso Amba
Sirinka
Fura
Bersu Terara
Gengoyta Terara
Gura'ale Terara
Damuta Terara
Mersa
Calloh
1980 m
Hamarko Terara
Uodedsca
Gelana
Wirgesa
Wichale
Dembel
Muoniu
Amba Sel
Hurmat Terara
Lake Hayk'
Legda
Boro
Hayk
As Aleyti Terara
Kuta Ber
Wereba Abo YeTerara Senselet
Boru
Sulula
Maghenta
Murtu
Guba YeTerara Senselet
Lake Hardibo
Bati
Uwa Terara
Motocolo
Dese
Garfa
Bifeta
Kombolcha
Hara YeTerara Senselet
Yegof Terara
Ancharo
Chedar Amba
Indode Terara
Harbu
Awash Wenz
Ieciuol
Kabe
1
2
2

29

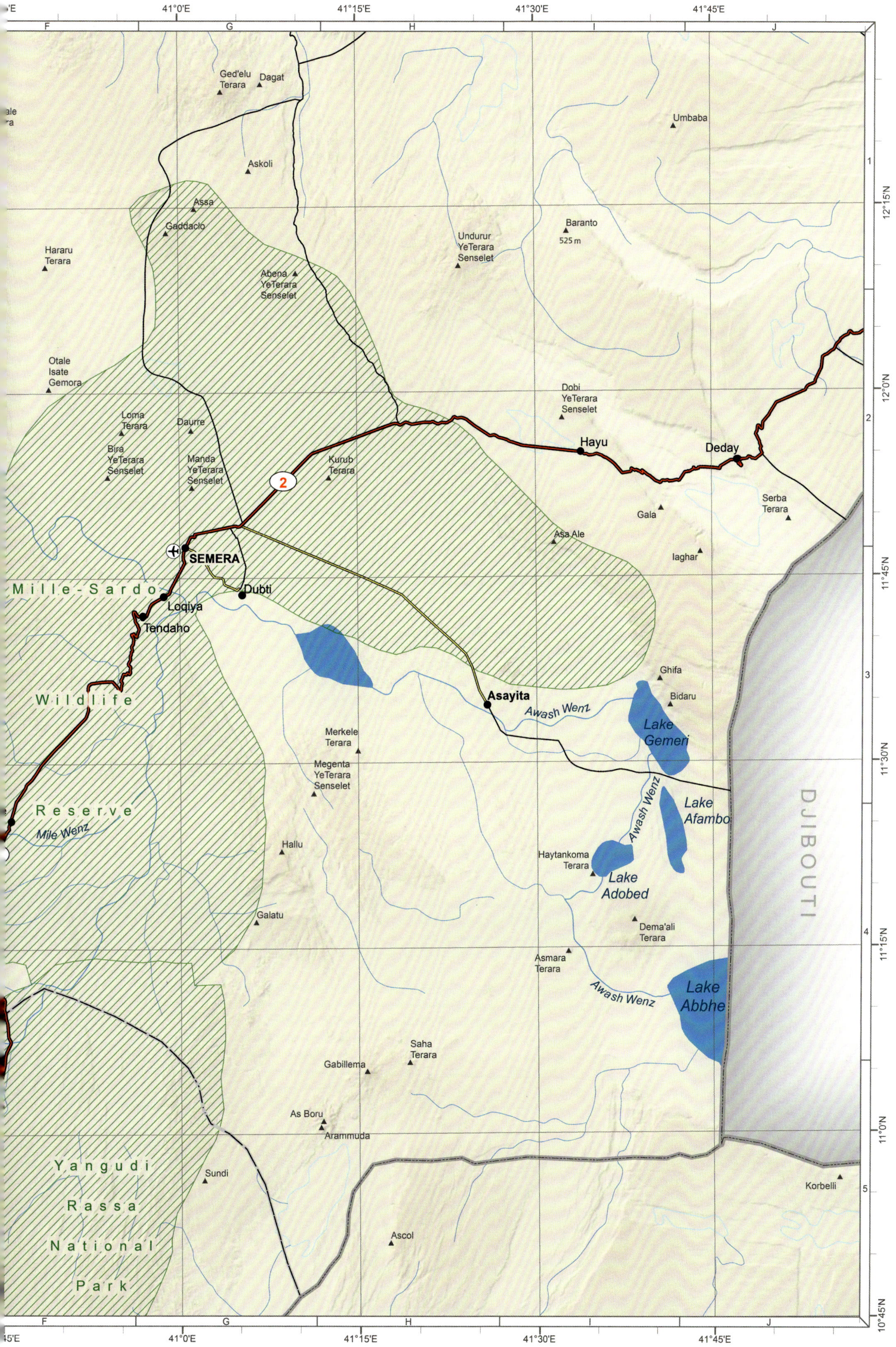
41°0'E
41°15'E
41°30'E
41°45'E
F
G
H
I
J
1
2
3
4
5
12°15'N
12°0'N
11°45'N
11°30'N
11°15'N
11°0'N
10°45'N
Ged'elu Terara
Dagat
Umbaba
Askoli
Assa
Gaddaclo
Hararu Terara
Undurur YeTerara Senselet
Baranto
525 m
Abena YeTerara Senselet
Otale Isate Gemora
Dobi YeTerara Senselet
Loma Terara
Daurre
Bira YeTerara Senselet
Manda YeTerara Senselet
Kurub Terara
2
Hayu
Deday
Serba Terara
Gala
Asa Ale
Iaghar
SEMERA
Dubti
Mille-Sardo
Loqiya
Tendaho
Wildlife
Reserve
Mile Wenz
Ghifa
Bidaru
Asayita
Awash Wenz
Lake Gemeri
Merkele Terara
Megenta YeTerara Senselet
Lake Afambo
DJIBOUTI
Hallu
Haytankoma Terara
Lake Adobed
Galatu
Dema'ali Terara
Asmara Terara
Lake Abbhe
Saha Terara
Gabillema
As Boru
Arammuda
Yangudi
Rassa
National
Park
Sundi
Korbelli
Ascol

30

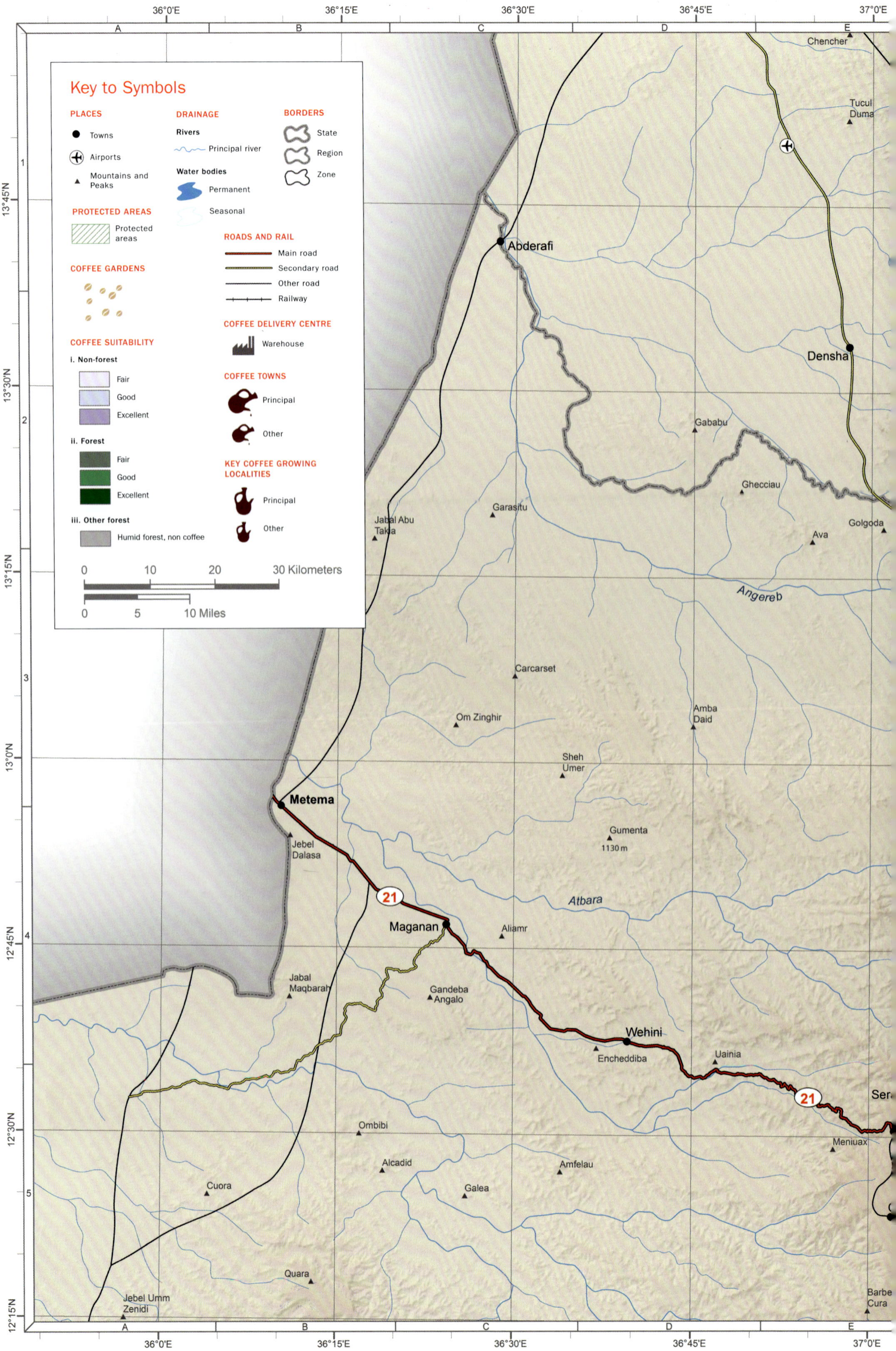
Key to Symbols
PLACES
Towns
Airports
Mountains and Peaks
PROTECTED AREAS
Protected areas
COFFEE GARDENS
COFFEE SUITABILITY
i. Non-forest
Fair
Good
Excellent
ii. Forest
Fair
Good
Excellent
iii. Other forest
Humid forest, non coffee
DRAINAGE
Rivers
Principal river
Water bodies
Permanent
Seasonal
BORDERS
State
Region
Zone
ROADS AND RAIL
Main road
Secondary road
Other road
Railway
COFFEE DELIVERY CENTRE
Warehouse
COFFEE TOWNS
Principal
Other
KEY COFFEE GROWING LOCALITIES
Principal
Other
0 10 20 30 Kilometers
0 5 10 Miles
36°0'E
36°15'E
36°30'E
36°45'E
37°0'E
13°45'N
13°30'N
13°15'N
13°0'N
12°45'N
12°30'N
12°15'N
A
B
C
D
E
1
2
3
4
5
Chencher
Tucul Duma
Abderafi
Densha
Gababu
Ghecciau
Garasitu
Jabal Abu Takla
Golgoda
Ava
Angereb
Carcarset
Om Zinghir
Amba Daid
Sheh Umer
Metema
Jebel Dalasa
Gumenta
1130 m
Atbara
21
Maganan
Aliamr
Jabal Maqbarah
Gandeba Angalo
Wehini
Encheddiba
Uainia
Ser
Meniuax
Ombibi
Alcadid
Amfelau
Cuora
Galea
Quara
Jebel Umm Zenidi
Barbe Cura
33

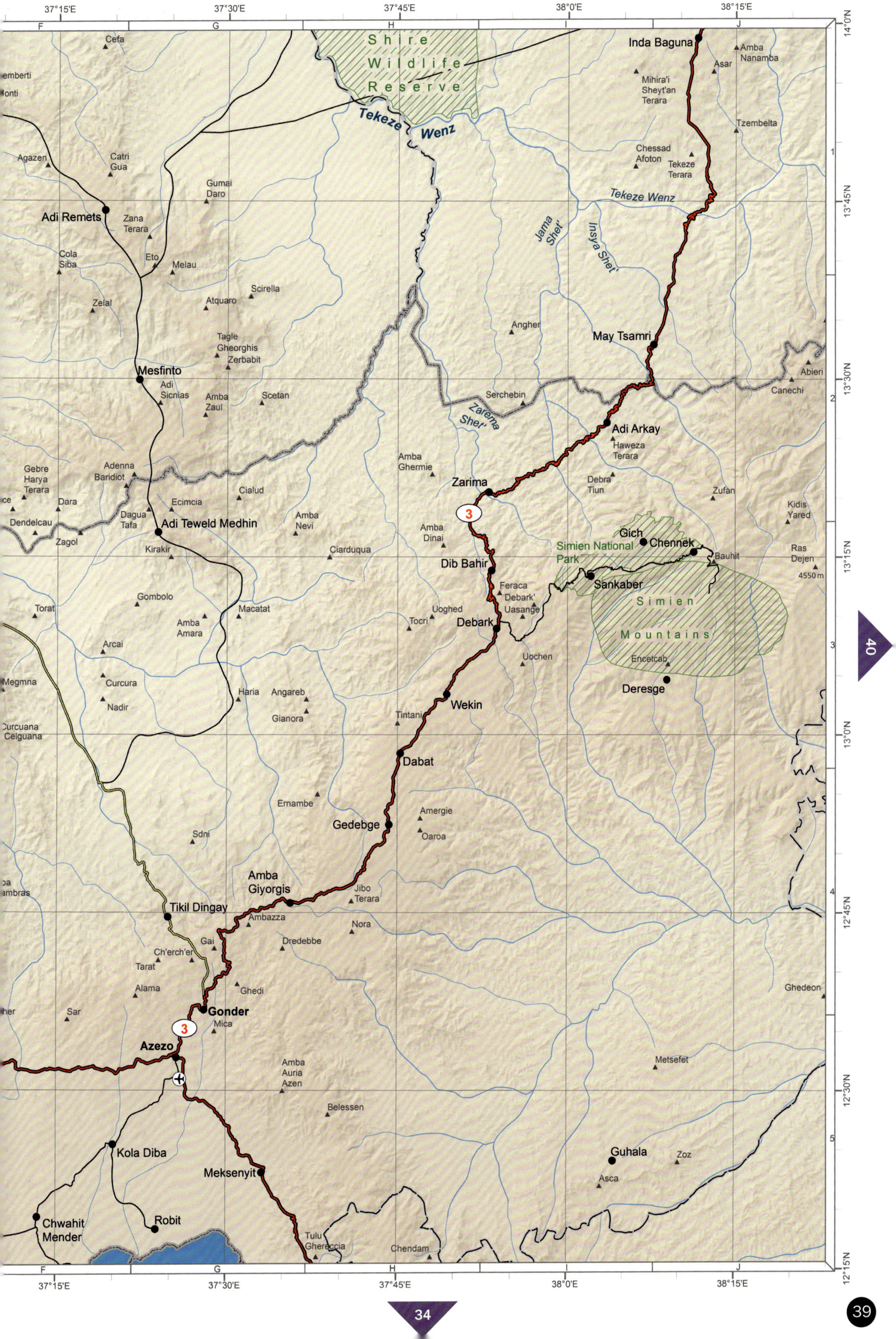

Shire Wildlife Reserve
Tekeze Wenz
Inda Baguna
Amba Nanamba
Asar
Mihira'i Sheyt'an Terara
Tzembelta
Chessad Afoton
Tekeze Terara
Adi Remets
Zana Terara
Cola Siba
Eto
Melau
Zelal
Atquaro
Scirella
Jama Shet'
Insya Shet'
Angher
May Tsamri
Mesfinto
Adi Sicnias
Amba Zaul
Scetan
Serchebin
Zarema Shet'
Adi Arkay
Haweza Terara
Abieri
Canechi
Amba Ghermie
Zarima
Debra Tiun
Zufan
Kidis Yared
Adi Teweld Medhin
Dib Bahir
Simien National Park
Gich
Chennek
Sankaber
Simien Mountains
Debark
Wekin
Dabat
Gedebge
Amba Giyorgis
Tikil Dingay
Gonder
Azezo
Kola Diba
Meksenyit
Chwahit Mender
Robit
Guhala
Deresge
Metsefet
Ghedeon

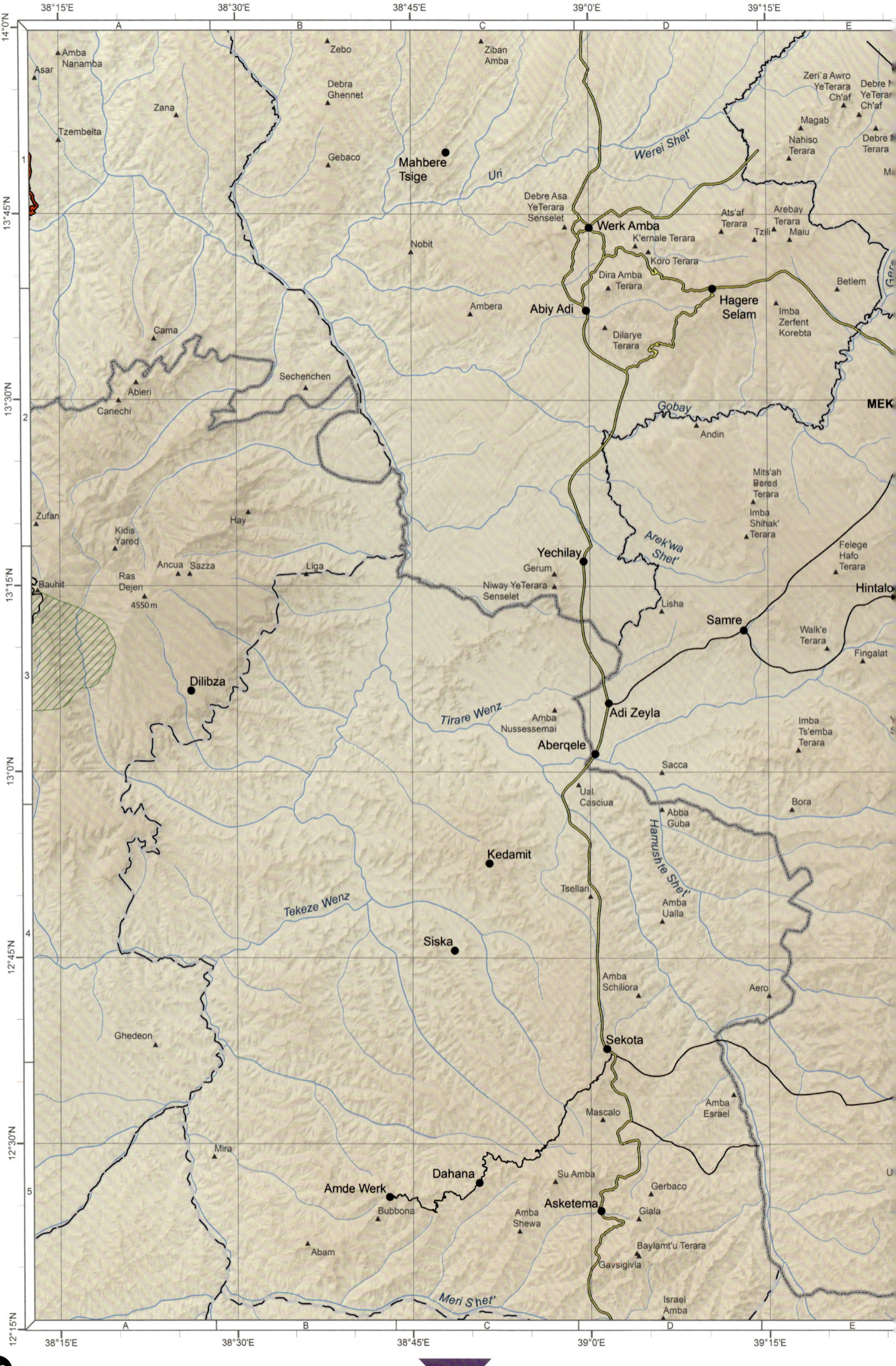

38°15'E
38°30'E
38°45'E
39°0'E
39°15'E
14°0'N
13°45'N
13°30'N
13°15'N
13°0'N
12°45'N
12°30'N
12°15'N
A
B
C
D
E
1
2
3
4
5
Amba Nanamba
Asar
Zebo
Ziban Amba
Debra Ghennet
Zana
Tzembelta
Gebaco
Mahbere Tsige
Uri
Werei Shet'
Zeri`a Awro YeTerara Ch'af
Magab
Nahiso Terara
Debre Asa YeTerara Senselet
Werk Amba
Ats'af Terara
Arebay Terara
Tzili
Maiu
K'ernale Terara
Nobit
Koro Terara
Dira Amba Terara
Hagere Selam
Betlem
Ambera
Abiy Adi
Imba Zerfent Korebta
Dilarye Terara
Cama
Sechenchen
Abieri
Canechi
Gobay
Andin
MEK
Mits'ah Bered Terara
Imba Shihak' Terara
Zufan
Hay
Kidis Yared
Arek'wa Shet'
Yechilay
Felege Hafo Terara
Gerum
Ancua
Sazza
Liga
Ras Dejen
4550 m
Bauhit
Niway YeTerara Senselet
Hintalo
Lisha
Samre
Walk'e Terara
Fingalat
Dilibza
Tirare Wenz
Amba Nussessemai
Adi Zeyla
Imba Ts'emba Terara
Aberqele
Sacca
Ual Casciua
Bora
Abba Guba
Hamushte Shet'
Kedamit
Tsellari
Amba Ualla
Tekeze Wenz
Siska
Amba Schiliora
Aero
Ghedeon
Sekota
Mascalo
Amba Esrael
Mira
Su Amba
Gerbaco
Dahana
Amde Werk
Bubbona
Asketema
Amba Shewa
Giala
Abam
Baylamt'u Terara
Gavsigivia
Meri S'het'
Israel Amba

39

35

39°45'E
40°0'E
40°15'E
40°30'E
40°45'E
14°0'N
13°45'N
13°30'N
13°15'N
13°0'N
12°45'N
12°30'N
12°15'N
F
G
H
I
J
1
2
3
4
5
Lake Karum
Lake Bakili
Mesk'el
Ts'arya Terara
Atsbi Inda Silase
Berahile
Chebrit Ale
Borele Ale
Wukro
Alu
Cabtai
Agula
Gabuli
Nahaso
Womberta
May Mekdan
Adai
Amaytoli Terara
Kwiha
Eghir Eriva
Barali
Garbena
Bolbala
Assavaha
Shehet
Wele'a Terara
Allemale
Adu Hill
Keyah
Buruole
Imba
Uogare
Uassi
Bit'bit' Terara
Hadele Terara
Ela Callo
Garagiam
Waga'ade Terara
Abbedda
Mahalta
Amba Dugghеducca
Afulu
Ghimmeri
Gumasir
Garule
Magalle
1690 m
Urikomam Terara
Amba Ferra
Adi Shoh
Limmo
Mehoni
Dabbaho
Haia Olel
Bar Ali
Halaite
Ada'al YeTerara Senselet
Rair
Tsoiag
Weyra Wuha
Hodale Terara
Gamabolu
Gomoyta YeTerara Senselet
Thec
Menda Terara
Waja

36

Fishing on Lake Awasa,
in the Great Rift Valley
Image: Jenny Williams

Gazetteer (geographical name index)

This gazetteer includes all the mapped geographical features included in the Atlas (apart from roads and rivers), placed into ten categories: zones; coffee-growing localities; coffee towns; coffee delivery centres; regional capitals; airports; protected areas; major lakes; summits/mountains; and towns/settlements. These names are referenced by Atlas map page number, alpha-numeric grid reference, region, zone and the latitude and longitude in degrees and minutes. The alpha-numeric grid reference refers to the letter, A to J (on the bottom of the facing pages) and the number 1 to 5 (on the side of the page).

Zones (administrative) are indicated by page number only, to provide an approximate indication of location; many of the zones span two or more pages. Page numbers and the specific names are given in bold, and zones are in italics. Regions are given in roman without bold or italics. In the gazetteer of towns/settlements, regional capitals are given in upper case (e.g. ASOSA).

Example

Page	Grid	Town	Region	Zone	Coordinates
10	D1	**Jimma**	Oromia	*Jimma*	7°40'N 36°50'E

The town of **Jimma** can be found in the Atlas on page **10**, at grid reference D1. The geographical coordinates 7°40'N 36°50'E can be used to locate this town to an accuracy of approximately 1km. Jimma is a town of the Oromia Region of the *Jimma* (administrative) zone.

Zones

Page	Zone	Region
7	***Afder***	Somali
8	***Agnuak***	Gambella
11	***Alaba***	SNNPR
21	***Arsi***	Oromia
24	***Asosa***	Benishangul Gumuz
33	***Awi/Agew***	Amhara
13	***Bale***	Oromia
3	***Basketo***	SNNPR
2	***Bench Maji***	SNNPR
4	***Borena***	Oromia
40	***Central***	Tigray
10	***Dawro***	SNNPR
30	***Dire Dawa***	Dire Dawa
Not in Atlas	***Doolo***	Somali
27	***East Gojam***	Amhara
22	***East Hararge***	Oromia
20	***East Shewa***	Oromia
18	***East Wellega***	Oromia
41	***Eastern***	Tigray
23	***Fafan***	Somali
4	***Gamo Gofa***	SNNPR
4	***Gedio (Gedeo)***	SNNPR
5	***Guji***	Oromia
19	***Gurage***	SNNPR
11	***Hadiya***	SNNPR
23	***Hareri***	Hareri
27	***Horo Guduru***	Oromia
17	***Ilubabor***	Oromia
23	***Jarar***	Somali
10	***Jimma***	Oromia
10	***Kaffa***	SNNPR
17	***Kelem Wellega***	Oromia
25	***Kemashi***	Benishangul Gumuz
10	***Konta***	SNNPR
Not in Atlas	***Korahe***	Somali
11	***Kembata Tembaro***	SNNPR
6	***Liben***	Somali
9	***Majang***	Gambella
32	***Metekel***	Benishangul Gumuz
15	***Nogob***	Somali
38	***North Gonder***	Amhara
28	***North Shewa (R3)***	Amhara
28	***North Shewa (R4)***	Oromia
39	***North Western***	Tigray
35	***North Wollo***	Amhara
16	***Nuer***	Gambella
29	***Oromia***	Amhara
20	***Region 14***	Addis Ababa
4	***Segen Peoples'***	SNNPR
11	***Selti***	SNNPR
15	***Shabelle***	Somali
9	***Sheka***	SNNPR
12	***Sidama***	SNNPR
30	***Siti***	Somali
34	***South Gonder***	Amhara
3	***South Omo***	SNNPR
19	***South West Shewa***	Oromia
28	***South Wollo***	Amhara
40	***Southern***	Tigray
36	***Special Woreda***	Amhara
40	***Wag Himra***	Amhara
12	***West Arsi***	Oromia
34	***West Gojam***	Amhara
22	***West Hararge***	Oromia
19	***West Shewa***	Oromia
17	***West Wellega***	Oromia
39	***Western***	Tigray
11	***Wolayita***	SNNPR
11	***Yem***	SNNPR
37	***Awsi (Zone 1)***	Afar
41	***Kilbati (Zone 2)***	Afar
29	***Gabi (Zone 3)***	Afar
36	***Fenti (Zone 4)***	Afar
29	***Hari (Zone 5)***	Afar

Coffee-growing localities

5	F2	**Guji**	Oromia	*Guji*	5°52'N 38°29'E
10	B3	**Kaffa**	SNNPR	*Kaffa*	7°15'N 36°7'E
18	D5	**Limu**	Oromia	*Jimma*	7°55'N 36°51'E
13	F5	**Bale (Harenna)**	Oromia	*Bale*	6°31'N 39°44'E
12	C5	**Bensa**	SNNPR	*Sidama*	6°29'N 38°51'E
12	B5	**Bona Zuria**	SNNPR	*Sidama*	6°30'N 38°42'E
10	C1	**Gera**	Oromia	*Jimma*	7°47'N 36°26'E
9	G4	**Gura Ferda**	SNNPR	*Bench Maji*	6°50'N 35°5'E
4	E2	**Kochere**	SNNPR	*Gedio*	6°5'N 38°13'E
9	G2	**Mezhenger Godere)**	Gambella	*Majang*	7°19'N 35°12'E
9	H3	**Sheko**	SNNPR	*Bench Maji*	7°5'N 35°23'E
12	D5	**W. Arsi (Nansebo)**	SNNPR	*Sidama*	6°28'N 39°5'E
12	A4	**Wonsho**	SNNPR	*Sidama*	6°43'N 38°28'E

Coffee towns

10	C1	**Agaro**	Oromia	*Jimma*	7°51'N 36°35'E
12	A4	**Aleta Wendo**	SNNPR	*Sidama*	6°36'N 38°25'E
10	B3	**Anderacha**	SNNPR	*Kaffa*	7°12'N 36°16'E
22	B2	**Asbe Teferi**	Oromia	*West Hararge*	9°4'N 40°51'E
9	H4	**Bebeka**	SNNPR	*Bench Maji*	6°51'N 35°20'E
22	E2	**Bedeno**	Oromia	*East Hararge*	9°6'N 41°37'E
4	B4	**Bekawile (Konso)**	SNNPR	*Segen Peoples'*	5°20'N 37°26'E
10	B3	**Bonga**	SNNPR	*Kaffa*	7°16'N 36°15'E
26	I2	**Debre Zeyit (Wembera)**	Benishangul Gumuz	*Metekel*	10°37'N 35°39'E
27	F1	**Finote Selam**	Amhara	*West Gojam*	10°40'N 37°15'E
16	E2	**Garigeda**	Oromia	*Kelem Wellega*	8°46'N 34°39'E
21	J2	**Gelemso**	Oromia	*West Hararge*	8°48'N 40°30'E
9	H5	**Gesha**	SNNPR	*Bench Maji*	6°33'N 35°30'E
17	J1	**Gimbi**	Oromia	*West Wellega*	9°10'N 35°50'E
21	H4	**Gololcha**	Oromia	*Arsi*	8°11'N 40°3'E
17	H4	**Gore**	Oromia	*Ilubabor*	8°9'N 35°32'E
22	C1	**Hirna**	Oromia	*West Hararge*	9°13'N 41°6'E
10	D1	**Jimma**	Oromia	*Jimma*	7°40'N 36°50'E
2	D1	**Maji**	SNNPR	*Bench Maji*	6°10'N 35°35'E
21	I3	**Mechara**	Oromia	*West Hararge*	8°36'N 40°20'E
22	C2	**Mesela**	Oromia	*West Hararge*	9°5'N 41°10'E
17	I4	**Metu**	Oromia	*Ilubabor*	8°18'N 35°34'E
9	I3	**Mizan Teferi**	SNNPR	*Bench Maji*	7°00'N 35°25'E
16	E3	**Sedi (Ashi)**	Oromia	*Kelem Wellega*	8°39'N 34°40'E
9	H2	**Tepi**	Gambella	*Majang*	7°12'N 35°25'E
4	E1	**Wenago**	Oromia	*Borena*	6°18'N 38°15'E
17	J4	**Yayu**	Oromia	*Ilubabor*	8°20'N 35°49'E
12	A4	**Yirga Alem**	SNNPR	*Sidama*	6°44'N 38°24'E
4	E1	**Yirgacheffe**	Oromia	*Borena*	6°9'N 38°11'E
34	B3	**Zege (Zeghie)**	Amhara	*West Gojam*	11°41'N 37°19'E

Coffee delivery centres

20	C2	**Addis Ababa**	Addis Ababa	*Region 14*	9°0'N 38°46'E
18	C3	**Bedele**	Oromia	*Ilubabor*	8°27'N 36°21'E
10	B3	**Bonga**	SNNPR	*Kaffa*	7°17'N 36°15'E
12	A5	**Dilla**	Oromia	*Borena*	6°24'N 38°18'E
31	F4	**Dire Dawa**	Dire Dawa	*Dire Dawa*	9°35'N 41°51'E
17	J1	**Gimbi**	Oromia	*West Wellega*	9°10'N 35°50'E
12	B3	**Hawassa**	SNNPR	*Sidama*	7°2'N 38°29'E
10	D1	**Jimma**	Oromia	*Jimma*	7°40'N 36°50'E
11	H4	**Soddo**	SNNPR	*Wolayita*	6°51'N 37°45'E

Regional capitals

20	C2	**Addis Ababa (Finfine)**	Addis Ababa/ Oromia	*Region 14*	9°1'N 38°45'E
24	E3	**Asosa**	Benishangul Gumuz	*Asosa*	10°3'N 34°32'E
34	B3	**Bahir Dar**	Amhara	*West Gojam*	11°35'N 37°22'E
31	F4	**Dire Dawa**	Dire Dawa	*Dire Dawa*	9°35'N 41°51'E
16	E4	**Gambela**	Gambella	*Agnuak*	8°15'N 34 °35'E
23	G1	**Harar**	Hareri	*Hareri*	9°18'N 42°7'E
12	B3	**Hawassa**	SNNPR	*Sidama*	7°2'N 38°29'E
23	I1	**Jijiga (Jigjiga)**	Somali	*Fafan*	9°20'N 42°47'E
40	E2	**Mekele**	Tigray	*Southern*	13°29'N 39°28'E
37	G3	**Semera**	Afar	*Awsi (Zone 1)*	11°47'N 41°0'E

Airports

20	C2	**Addis Ababa**	Addis Ababa	*Region 14*	8°58'N 38°47'E
4	C2	**Arba Minch**	SNNPR	*Gamo Gofa*	6°2'N 37°35'E
24	E3	**Assosa**	Benishangul Gumuz	*Asosa*	10°1'N 34°34'E
34	B3	**Bahir Dar**	Amhara	*West Gojam*	11°36'N 37°19'E
31	F4	**Dire Dawa**	Dire Dawa	*Dire Dawa*	9°37'N 41°51'E
16	E4	**Gambela**	Gambella	*Agnuak*	8°7'N 34°33'E
39	G5	**Gonder**	Amhara	*North Gonder*	12°30'N 37°25'E
38	E1	**Humera**	Tigray	*Western*	13°49'N 36°54'E
23	J1	**Jijiga**	Somali	*Fafan*	9°19'N 42°54'E
10	D1	**Jimma**	Oromia	*Jimma*	7°40'N 36°48'E
35	H2	**Lalibela**	Amhara	*North Wollo*	11°58'N 38°59'E
41	F2	**Mekele**	Tigray	*Southern*	13°28'N 39°31'E
13	H3	**Robe (Goba)**	Oromia	*Bale*	7°7'N 40°2'E
37	G3	**Semera**	Afar	*Awsi (Zone 1)*	11°47'N 40°59'E

Protected areas

19	G5	**Abelti Gibie National Forest Priority Area**	Oromia	*Jimma*	8°3'N 37°34'E
12	B2	**Abijatta-Shalla Lakes National Park**	Oromia	*West Arsi*	7°32'N 38°31'E
21	I1	**Alledeghi Wildlife Reserve**	Oromia	*West Hararge*	9°10'N 40°26'E
21	H2	**Awash National Park**	Oromia	*East Shewa*	8°54'N 40°2'E
21	G2	**Awash West Wildlife Reserve**	Oromia	*East Shewa*	8°58'N 39°51'E
13	F4	**Bale Wildlife Reserve**	Oromia	*West Arsi*	6°55'N 39°43'E
13	F4	**Bale Mountains National Park**	Oromia	*West Arsi*	6°55'N 39°40'E
3	I5	**Chelbi Wildlife Reserve**	SNNPR	*South Omo*	5°00'N 36°47'E
16	C5	**Gambella National Park**	Gambella	*Agnuak*	7°57'N 34°7'E
30	A2	**Gewane Wildlife Reserve**	Afar	*Gabi (Zone 3)*	10°24'N 40°42'E
10	A2	**Kaffa UNESCO-MAB Biosphere Reserve**	SNNPR	*Kaffa*	7°25'N 35°56'E
3	G3	**Mago National Park**	SNNPR	*South Omo*	5°32'N 36°15'E
37	G3	**Mille-Sardo Wildlife Reserve**	Afar	*Awsi (Zone 1)*	11°41'N 40°59'E
4	D2	**Nechisar National Park**	Oromia	*Borena*	5°59'N 37°53'E
2	E2	**Omo National Park**	SNNPR	*Bench Maji*	5°55'N 35°45'E
39	H1	**Shire Wildlife Reserve**	Tigray	*North Western*	13°56'N 37°45'E
39	J3	**Simien Mountains**	Amhara	*North Gonder*	13°9'N 38°10'E
39	I3	**Simien National Park**	Amhara	*North Gonder*	13°15'N 38°5'E
3	F2	**Tama Wildlife Reserve**	SNNPR	*South Omo*	5°57'N 36°4'E
5	G5	**Yabello Wildlife Sanctuary**	Oromia	*Borena*	4°47'N 38°30'
37	F5	**Yangudi Rassa National Park**	Afar	*Gabi (Zone 3)*	10°57'N 40°46'E
17	J4	**Yayu UNESCO-MAB Biosphere Reserve**	Oromia	*Ilubabor*	8°26'N 35°48'E

Major lakes

20	B2	**Aba-Samuel Reservoir**	Oromia	*South West Shewa*	8°49'N 38°42'E
20	A5	**Abay Hayk'**	SNNPR	*Selti*	7°56'N 38°21'E
19	I2	**Dendi Hayk'**	Oromia	*West Shewa*	8°50'N 38°0'E
27	F5	**Fincha-a Reservoir**	Oromia	*Horo Guduru*	9°28'N 37°12'E
11	H5	**Lake Abaya**	SNNPR	*Gamo Gofa*	6°20'N 37°50'E
37	J4	**Lake Abbhe**	Afar	*Awsi (Zone 1)*	11°9'N 41°45'E
12	B2	**Lake Abiyala**	Oromia	*West Arsi*	7°36'N 38°35'E
37	I4	**Lake Adobed**	Afar	*Awsi (Zone 1)*	11°22'N 41°36'E
37	I4	**Lake Afambo**	Afar	*Awsi (Zone 1)*	11°24'N 41°41'E
12	A3	**Lake Awasa (Hawassa)**	SNNPR	*Sidama*	7°3'N 38°26'E
29	I4	**Lake Bada**	Afar	*Gabi (Zone 3)*	9°54'N 40°23'E
41	I1	**Lake Bakili**	Afar	*Kilbati (Zone 2)*	13°56'N 40°30'E
21	G2	**Lake Basaka**	Oromia	*East Shewa*	8°52'N 39°52'E
4	B2	**Lake Chamo**	SNNPR	*Gamo Gofa*	5°50'N 37°33'E
20	D3	**Lake Chelekleka**	Oromia	*East Shewa*	8°46'N 38°58'E
12	B3	**Lake Dabashi**	SNNPR	*Sidama*	7°5'N 38°33'E
37	I3	**Lake Gemeri**	Afar	*Awsi (Zone 1)*	11°32'N 41°40'E
36	B4	**Lake Hardibo**	Amhara	*South Wollo*	11°14'N 39°46'E

36	B4	**Lake Hayk**	Amhara	*South Wollo*	11°21'N 39°43'E
41	I1	**Lake Karum (Assale)**	Afar	*Kilbati (Zone 2)*	13°59'N 40°27'E
20	D4	**Lake Koka**	Oromia	*East Shewa*	8°23'N 39°4'E
12	C2	**Lake Langano**	Oromia	*West Arsi*	7°35'N 38°45'E
29	H5	**Lake Liado**	Afar	*Gabi (Zone 3)*	9°33'N 40°12'E
12	B2	**Lake Shala (Shalla)**	Oromia	*West Arsi*	7°27'N 38°30'E
34	B2	**Lake Tana**	Amhara	*West Gojam/ South Gonder/ North Gonder*	12°0'N 37°18'E
20	C5	**Lake Ziway**	Oromia	*East Shewa*	7°59'N 38°50'E
29	I3	**Ozero Yard Khayk**	Afar	*Gabi (Zone 3)*	10°13'N 40°29'E
19	I3	**Wenchi'i Hayk'**	Oromia	*South West Shewa*	8°47'N 37°53'E

Summits/Mountains

A

29	G2	**Aba Boker Terara**	Amhara	*North Shewa (R3)*	10°26'N 39°54'E
18	C2	**Abalo**	Oromia	*East Wellega*	9°7'N 36°23'E
35	G1	**Abam**	Amhara	*Wag Himra*	12°22'N 38°36'E
12	B3	**Abaro**	Oromia	*West Arsi*	7°7'N 38°38'E
40	D4	**Abba Guba**	Amhara	*Wag Himra*	12°57'N 39°6'E
41	F3	**Abbedda**	Tigray	*Southern*	13°5'N 39°45'E
21	I5	**Abdoy Terara**	Oromia	*Arsi*	8°0'N 40°17'E
37	G1	**Abena YeTerara Senselet**	Afar	*Awsi (Zone 1)*	12°10'N 41°10'E
22	C2	**Abenyo Terara**	Oromia	*West Hararge*	8°57'N 41°9'E
40	J2	**Abieri**	Amhara	*North Gonder*	13°31'N 38°21'E
13	J2	**Abinas**	Oromia	*Bale*	7°29'N 40°30'E
13	I2	**Abol K'asim Terara**	Oromia	*Bale*	7°31'N 40°29'E
34	B4	**Abola**	Amhara	*West Gojam*	11°18'N 37°31'E
33	I1	**Abracioc**	Amhara	*North Gonder*	12°12'N 36°53'E
32	B3	**Abu Ramla**	Benishangul Gumuz	*Metekel*	11°41'N 35°7'E
13	J1	**Abul Casim**	Oromia	*Arsi*	7°43'N 40°31'E
35	I1	**Abune Yosef**	Amhara	*North Wollo*	12°9'N 39°11'E
26	C3	**Abutalla**	Oromia	*East Wellega*	10°10'N 36°25'E
41	J5	**Ada'al YeTerara Senselet**	Afar	*Awsi (Zone 1)*	12°36'N 40°44'E
41	F2	**Adai**	Tigray	*Southern*	13°33'N 39°43'E
22	C1	**Adem Gadi Burk'a Terara**	Oromia	*East Hararge*	9°9'N 41°16'E
41	F1	**Ademesso**	Tigray	*Eastern*	13°46'N 39°33'E
21	H4	**Ademu Terara**	Oromia	Arsi	8°21'N 40°2'E
39	G2	**Adenna**	Tigray	*Western*	13°22'N 37°22'E
35	I2	**Adeno**	Amhara	*North Wollo*	12°5'N 39°8'E
35	H2	**Aderbo**	Amhara	*North Wollo*	12°7'N 39°2'E
10	B3	**Adi**	SNNPR	*Kaffa*	7°2'N 36°9'E
35	I2	**Adi Dega Amba**	Amhara	*North Wollo*	12°5'N 39°6'E
39	G2	**Adi Sicnias**	Tigray	*Western*	13°28'N 37°24'E
41	H2	**Adu Hill**	Afar	*Kilbati (Zone 2)*	13°19'N 40°3'E
19	F1	**Adulan Terara**	Oromia	*West Shewa*	9°13'N 37°10'E
40	E4	**Aero**	Amhara	*Wag Himra*	12°42'N 39°15'E
30	A5	**Afdem Terara**	Somali	*Siti*	9°30'N 40°50'E
16	E4	**Afillu**	Oromia	*Kelem Wellega*	8°25'N 34°39'E
41	H3	**Afulu**	Afar	*Kilbati (Zone 2)*	12°57'N 40°11'E
13	H4	**Agal Terara**	Oromia	*Bale*	6°57'N 40°8'E
10	D4	**Agarri**	SNNPR	*Dawro*	6°55'N 36°50'E
39	F1	**Agazen**	Tigray	*Western*	13°48'N 37°14'E
10	C3	**Aggio**	SNNPR	*Kaffa*	7°9'N 36°37'E
30	D5	**Ahmar Mountains**	Oromia	*East Hararge*	9°25'N 41°26'E
33	H4	**Ailungiucsa**	Amhara	*Awi/Agew*	11°10'N 36°36'E
5	J3	**Aiso**	Oromia	*Guji*	5°29'N 39°26'E
31	I1	**Akaley**	Somali	*Siti*	10°58'N 42°42'E
22	E1	**Ala**	Oromia	*East Hararge*	9°14'N 41°42'E
26	C4	**Alaba Meda**	Oromia	*East Wellega*	9°54'N 36°35'E
12	B3	**Alacha**	Oromia	*West Arsi*	7°7'N 38°34'E
39	G4	**Alama**	Amhara	*North Gonder*	12°38'N 37°22'E
9	I2	**Alamu**	SNNPR	*Kaffa*	7°23'N 35°35'E
8	E4	**Alarra**	Gambella	*Agnuak*	6°49'N 34°37'E
38	B5	**Alcadid**	Amhara	*North Gonder*	12°27'N 36°19'E
19	G4	**Ali**	Oromia	*Jimma*	8°11'N 37°35'E
38	C4	**Aliamr**	Amhara	*North Gonder*	12°46'N 36°29'E
41	F2	**Allemale**	Tigray	*Southern*	13°20'N 39°35'E
41	H1	**Alu**	Afar	*Kilbati (Zone 2)*	13°46'N 40°14'E
9	I3	**Aman Mountains**	SNNPR	*Bench Maji*	7°1'N 35°34'E
3	I3	**Amar**	SNNPR	*South Omo*	5°35'N 36°45'E
31	G4	**Amara Terara**	Dire Dawa	*Dire Dawa*	9°45'N 42°13'E
41	J2	**Amaytoli Terara**	Afar	*Kilbati (Zone 2)*	13°31'N 40°38'E
41	F3	**Amba Alage**	Tigray	*Southern*	12°59'N 39°33'E
39	G3	**Amba Amara**	Amhara	*North Gonder*	13°10'N 37°28'E
41	F2	**Amba Aradam**	Tigray	*Southern*	13°20'N 39°31'E
39	G5	**Amba Auria Azen**	Amhara	*North Gonder*	12°30'N 37°35'E
39	F4	**Amba Balambras**	Amhara	*North Gonder*	12°46'N 37°9'E
39	F2	**Amba Cucoi**	Amhara	*North Gonder*	13°18'N 37°6'E
38	D3	**Amba Daid**	Amhara	*North Gonder*	13°3'N 36°45'E
39	H3	**Amba Dinai**	Amhara	*North Gonder*	13°16'N 37°49'E
41	G3	**Amba Duggheducca**	Tigray	*Southern*	12°58'N 39°46'E
40	D5	**Amba Esrael**	Amhara	*Wag Himra*	12°34'N 39°12'E
35	H5	**Amba Farit**	Amhara	*South Wollo*	10°56'N 38°58'E
41	F4	**Amba Ferra**	Tigray	*Southern*	12°52'N 39°39'E
39	H2	**Amba Ghermie**	Amhara	*North Gonder*	13°22'N 37°48'E
39	F1	**Amba Mahar**	Tigray	*Western*	13°43'N 37°6'E
35	J3	**Amba Moka**	Amhara	*South Wollo*	11°31'N 39°22'E
39	J1	**Amba Nanamba**	Tigray	*North Western*	13°58'N 38°15'E
39	G3	**Amba Nevi**	Amhara	*North Gonder*	13°17'N 37°36'E
40	C3	**Amba Nussessemai**	Amhara	*Wag Himra*	13°5'N 38°57'E
40	D4	**Amba Schiliora**	Amhara	*Wag Himra*	12°42'N 39°4'E
36	B4	**Amba Sel**	Amhara	*South Wollo*	11°27'N 39°48'E
40	C5	**Amba Shewa**	Amhara	*Wag Himra*	12°23'N 38°54'E
40	D4	**Amba Ualla**	Amhara	*Wag Himra*	12°48'N 39°6'E
39	G2	**Amba Zaul**	Tigray	*Western*	13°27'N 37°28'E
39	G4	**Ambazza**	Amhara	*North Gonder*	12°44'N 37°32'E
40	C2	**Ambera**	Tigray	*Central*	13°37'N 38°50'E
11	H3	**Ambericho Terara**	SNNPR	*Kembata Tembaro*	7°17'N 37°52'E
34	B4	**Amedamit**	Amhara	*West Gojam*	11°10'N 37°25'E
39	H4	**Amergie**	Amhara	*North Gonder*	12°53'N 37°47'E
38	C5	**Amfelau**	Amhara	*North Gonder*	12°27'N 36°34'E
30	A3	**Amoissa**	Afar	*Gabi (Zone 3)*	10°4'N 40°49'E
34	E5	**Amoneuos**	Amhara	*East Gojam*	10°58'N 38°16'E
27	H2	**Amora**	Amhara	*East Gojam*	10°35'N 37°46'E
34	D2	**Amora Gedel**	Amhara	*South Gonder*	12°1'N 37°49'E
9	H3	**Amora Gedel Terara**	SNNPR	*Bench Maji*	7°7'N 35°29'E
21	F4	**Amut'a Terara**	Oromia	*Arsi*	8°9'N 39°34'E
20	E5	**Ancfola**	Oromia	*Arsi*	8°6'N 39°28'E
39	F5	**Ancher**	Amhara	*North Gonder*	12°36'N 37°9'E
40	A3	**Ancua (Weynobar)**	Amhara	*North Gonder*	13°16'N 38°25'E
40	D2	**Andin**	Tigray	*Southern*	13°28'N 39°9'E
9	H2	**Andracha YeTerara Senselet**	SNNPR	*Sheka*	7°30'N 35°30'E
38	C4	**Angalo**	Amhara	*North Gonder*	12°41'N 36°23'E
39	H3	**Angareb**	Amhara	*North Gonder*	13°3'N 37°37'E
39	I2	**Angher**	Tigray	*North Western*	13°34'N 37°55'E
21	G1	**Ankober**	Amhara	*North Shewa (R3)*	9°18'N 39°45'E
20	C3	**Anole Terara**	Oromia	*East Shewa*	8°29'N 38°51'E
31	F1	**Ara le Koma**	Somali	*Siti*	10°58'N 42°7'E
23	F2	**Arab**	Oromia	*East Hararge*	8°59'N 41°54'E
31	G1	**Arabdara**	Somali	*Siti*	10°54'N 42°13'E
13	J3	**Arabe YeTerara Senselet**	Oromia	*Bale*	7°3'N 40°32'E
13	I2	**Arablig**	Oromia	*Bale*	7°21'N 40°28'E
31	I1	**Arag**	Somali	*Siti*	10°52'N 42°42'E
37	H5	**Arammuda**	Afar	*Awsi (Zone 1)*	11°1'N 41°12'E
12	D4	**Aranna**	Oromia	*West Arsi*	6°40'N 39°0'E
27	H2	**Arat Mekera Kir**	Amhara	*East Gojam*	10°36'N 37°45'E
18	E5	**Araua**	Oromia	*Jimma*	8°6'N 37°4'E
21	I3	**Arba Gugu Terara**	Oromia	*West Hararge*	8°40'N 40°15'E
39	F3	**Arcai**	Amhara	*North Gonder*	13°7'N 37°19'E
7	F1	**Arce**	Oromia	*Bale*	6°10'N 40°40'E
29	I2	**Ardgera**	Afar	*Hari (Zone 5)*	10°34'N 40°16'E
40	E1	**Arebay Terara**	Tigray	*Central*	13°44'N 39°16'E
3	H2	**Argun**	SNNPR	*South Omo*	5°59'N 36°40'E
18	C2	**Arjo**	Oromia	*East Wellega*	8°49'N 36°28'E
26	C5	**Ark'umbe Terara**	Oromia	*East Wellega*	9°32'N 36°28'E
31	H2	**Arome**	Somali	*Siti*	10°22'N 42°35'E
6	B3	**Arsi Cormoso**	Oromia	*Bale*	5°38'N 39°45'E
36	D4	**As Aleyti Terara**	Afar	*Awsi (Zone 1)*	11°18'N 40°11'E
37	H5	**As Boru**	Afar	*Awsi (Zone 1)*	11°1'N 41°12'E
37	I2	**Asa Ale**	Afar	*Awsi (Zone 1)*	11°48'N 41°32'E
29	H4	**Asa Kuma Terara**	Afar	*Hari (Zone 5)*	9°53'N 40°4'E
30	C2	**As'ali Terara**	Somali	*Siti*	10°33'N 41°11'E
39	J1	**Asar**	Tigray	*North Western*	13°56'N 38°13'E

34	E1	**Asca**	Amhara	*North Gonder*	12°22'N 38°3'E
30	C1	**Ascol**	Somali	*Siti*	10°51'N 41°18'E
21	J1	**Asebot Terara**	Oromia	*West Hararge*	9°16'N 40°34'E
31	H1	**Asengedi**	Somali	*Siti*	10°58'N 42°34'E
34	C2	**Asiba**	Amhara	*South Gonder*	12°7'N 37°47'E
29	I4	**Asi'ole Terara**	Afar	*Gabi (Zone 3)*	9°50'N 40°20'E
37	G1	**Askoli**	Afar	*Awsi (Zone 1)*	12°18'N 41°6'E
37	I4	**Asmara Terara**	Afar	*Awsi (Zone 1)*	11°15'N 41°33'E
31	H1	**Asmarerle**	Somali	*Siti*	10°57'N 42°38'E
36	E3	**Asoklo YeTerara Senselet**	Afar	*Awsi (Zone 1)*	11°34'N 40°40'E
37	G1	**Assa**	Afar	*Awsi (Zone 1)*	12°15'N 41°1'E
41	H2	**Assavaha**	Afar	*Kilbati (Zone 2)*	13°22'N 40°3'E
33	J5	**Assem Sellasie**	Amhara	*Awi/Agew*	11°0'N 37°0'E
39	G2	**Atquaro**	Tigray	*Western*	13°36'N 37°28'E
40	D1	**Ats'af Terara**	Tigray	*Central*	13°44'N 39°11'E
18	D4	**Auangero**	Oromia	*Jimma*	8°19'N 36°47'E
14	C3	**Auatu**	Oromia	*Bale*	7°18'N 41°6'E
7	J1	**Audo Range**	Somali	*Afder*	6°10'N 41°45'E
38	E2	**Ava**	Amhara	*North Gonder*	13°18'N 36°55'E
30	A3	**Ayelu Terara**	Afar	*Gabi (Zone 3)*	10°5'N 40°42'E
13	F1	**Azama**	Oromia	*Arsi*	7°49'N 39°38'E
34	E2	**Azer**	Amhara	*South Gonder*	12°5'N 38°8'E

B

20	F5	**Bada**	Oromia	*Arsi*	7°55'N 39°23'E
19	H3	**Badda Rogghie**	Oromia	*West Shewa*	8°43'N 37°41'E
18	E3	**Badda Ummo**	Oromia	*Jimma*	8°30'N 36°54'E
33	F2	**Bafa**	Amhara	*Awi/Agew*	11°56'N 36°9'E
2	B1	**Bai**	SNNPR	*Bench Maji*	6°16'N 35°11'E
10	C4	**Baka**	SNNPR	*Konta*	6°49'N 36°35'E
12	A1	**Balch'i**	SNNPR	*Selti*	7°50'N 38°20'E
3	A5	**Baldo**	Oromia	*Borena*	4°47'N 37°6'E
23	I3	**Bal'-Dzhi**	Oromia	*East Hararge*	8°37'N 42°42'E
31	I5	**Bale Terara**	Somali	*Fafan*	9°23'N 42°41'E
24	E4	**Bambesi**	Benishangul Gumuz	*Asosa*	9°43'N 34°40'E
32	C4	**Bameza**	Benishangul Gumuz	*Metekel*	11°10'N 35°14'E
23	I3	**Ban-Askulle**	Oromia	*East Hararge*	8°31'N 42°49'E
2	D1	**Bangal**	SNNPR	*Bench Maji*	6°11'N 35°33'E
41	I4	**Bar Ali**	Afar	*Fenti (Zone 4)*	12°37'N 40°23'E
41	I2	**Barali**	Afar	*Kilbati (Zone 2)*	13°27'N 40°19'E
12	E4	**Baranta Terara**	Oromia	*West Arsi*	6°47'N 39°27'E
37	I1	**Baranto**	Afar	*Awsi (Zone 1)*	12°13'N 41°33'E
33	J1	**Barberi Cura**	Amhara	*North Gonder*	12°16'N 37°0'E
2	C2	**Barchu**	SNNPR	*Bench Maji*	5°54'N 35°19'E
33	G2	**Bardanchi**	Amhara	*Awi/Agew*	12°0'N 36°21'E
39	F2	**Baridiot**	Tigray	*Western*	13°21'N 37°21'E
7	J1	**Basra**	Somali	*Afder*	6°10'N 41°44'E
13	F4	**Batu**	Oromia	*Bale*	6°55'N 39°44'E
39	J3	**Bauhit (Bwahit)**	Amhara	*North Gonder*	13°15'N 38°13'E
35	H1	**Baylamt'u Terara**	Amhara	*Wag Himra*	12°21'N 39°4'E
19	I1	**Bedat'ina Amba**	Oromia	*West Shewa*	9°18'N 37°56'E
33	G3	**Belaya**	Benishangul Gumuz	*Metekel*	11°28'N 36°12'E
18	E1	**Belbela YeTerara Ch'af**	Oromia	*East Wellega*	9°21'N 37°4'E
21	I3	**Belbeleyti YeTerara Senselet**	Oromia	*West Hararge*	8°48'N 40°27'E
24	E4	**Belbiso**	Benishangul Gumuz	*Asosa*	9°52'N 34°37'E
24	D2	**Beldoesu**	Benishangul Gumuz	*Asosa*	10°19'N 34°23'E
39	H5	**Belessen**	Amhara	*North Gonder*	12°28'N 37°39'E
18	E4	**Belida**	Oromia	*Jimma*	8°24'N 36°58'E
3	H2	**Belio**	SNNPR	*South Omo*	5°54'N 36°30'E
9	F2	**Belo**	Gambella	*Agnuak*	7°22'N 34°51'E
18	A2	**Bendi**	Oromia	*West Wellega*	8°57'N 36°0'E
10	C4	**Benkira**	SNNPR	*Konta*	6°39'N 36°29'E
5	H2	**Berbissa Tuto**	Oromia	*Guji*	6°1'N 38°53'E
20	D1	**Berek Terara**	Oromia	*North Shewa (R4)*	9°13'N 39°0'E
25	G1	**Beri**	Benishangul Gumuz	*Metekel*	10°48'N 35°14'E
21	F3	**Bericha Terara**	Oromia	*East Shewa*	8°37'N 39°31'E
36	E3	**Bersu Terara**	Afar	*Awsi (Zone 1)*	11°44'N 40°27'E
9	F2	**Betagheue**	Gambella	*Agnuak*	7°20'N 34°53'E
40	E1	**Betlem**	Tigray	*Central*	13°39'N 39°21'E
19	F2	**Betuma**	Oromia	*West Shewa*	9°4'N 37°12'E
24	E3	**Bibbio**	Benishangul Gumuz	*Asosa*	10°14'N 34°44'E
37	I3	**Bidaru**	Afar	*Awsi (Zone 1)*	11°35'N 41°41'E
13	F2	**Bidu Bala**	Oromia	*Arsi*	7°30'N 39°40'E
36	C5	**Bifeta**	Amhara	*Oromia*	11°5'N 40°4'E
19	H2	**Bifti YeTerara Ch'af**	Oromia	*West Shewa*	9°1'N 37°43'E
22	A5	**Bilik'a Terara**	Oromia	*West Hararge*	8°2'N 40°49'E
37	F2	**Bira YeTerara Senselet**	Afar	*Awsi (Zone 1)*	11°53'N 40°54'E
23	I2	**Birdid**	Oromia	*East Hararge*	8°48'N 42°46'E
41	F3	**Bit'bit'Terara**	Tigray	*Southern*	13°8'N 39°35'E
11	H2	**Bobigia**	SNNPR	*Hadiya*	7°23'N 37°44'E
18	E3	**Boca**	Oromia	*Jimma*	8°35'N 37°5'E
20	E2	**Boccan**	Amhara	*North Shewa (R3)*	8°53'N 39°16'E
9	G4	**Bocol**	Gambella	*Agnuak*	6°42'N 35°6'E
3	G3	**Bodu**	SNNPR	*South Omo*	5°46'N 36°14'E
18	B4	**Boia**	Oromia	*Ilubabor*	8°17'N 36°20'E
19	H2	**Bola**	Oromia	*West Shewa*	8°49'N 37°42'E
19	G2	**Bola Rogghie**	Oromia	*West Shewa*	8°54'N 37°37'E
41	F2	**Bolbala**	Tigray	*Southern*	13°23'N 39°38'E
19	I2	**Bole Terara**	Oromia	*West Shewa*	9°2'N 38°4'E
11	G3	**Bolosso**	SNNPR	*Wolayita*	7°3'N 37°36'E
32	C3	**Bongo**	Benishangul Gumuz	*Metekel*	11°32'N 35°15'E
12	C5	**Bonsa Terara**	SNNPR	*Sidama*	6°29'N 38°47'E
40	E4	**Bora**	Tigray	*Southern*	12°57'N 39°17'E
26	E4	**Bora Terara**	Oromia	*East Wellega*	9°56'N 36°53'E
30	E4	**Bora'at Terara**	Somali	*Siti*	9°46'N 41°39'E
26	D2	**Boramu**	Amhara	*West Gojam*	10°25'N 36°42'E
10	E1	**Bore**	Oromia	*Jimma*	7°52'N 37°2'E
41	J1	**Borele Ale**	Afar	*Kilbati (Zone 2)*	13°48'N 40°33'E
36	A4	**Boro**	Amhara	*South Wollo*	11°22'N 39°39'E
30	B3	**Boroli**	Somali	*Siti*	10°7'N 41°2'E
21	F4	**Borora**	Oromia	*Arsi*	8°15'N 39°32'E
20	D4	**Borra**	Oromia	*Arsi*	8°16'N 39°2'E
20	E3	**Boru T'oru Terara**	Oromia	*East Shewa*	8°43'N 39°15'E
20	E3	**Boset Terara**	Oromia	*East Shewa*	8°33'N 39°29'E
17	I5	**Bota**	Oromia	*Ilubabor*	7°54'N 35°39'E
19	I2	**Boti**	Oromia	*West Shewa*	8°51'N 38°3'E
19	F5	**Botor**	Oromia	*Jimma*	8°6'N 37°12'E
40	B5	**Bubbona**	Amhara	*Wag Himra*	12°24'N 38°42'E
4	A5	**Bube**	Oromia	*Borena*	4°49'N 37°11'E
11	J2	**Bubisa**	SNNPR	*Alaba*	7°24'N 38°14'E
26	B1	**Bulan**	Benishangul Gumuz	*Metekel*	10°41'N 36°6'E
31	J4	**Bulloynko**	Oromia	*East Hararge*	9°38'N 42°54'E
18	C4	**Bunno**	Oromia	*Ilubabor*	8°20'N 36°30'E
31	G2	**Bur Degamedou Dzher**	Somali	*Siti*	10°33'N 42°17'E
31	H2	**Bur Madu**	Somali	*Siti*	10°24'N 42°29'E
20	C1	**Bura Terara**	Oromia	*North Shewa (R4)*	9°16'N 38°51'E
31	I1	**Burale**	Somali	*Siti*	10°58'N 42°51'E
3	I4	**Burie**	SNNPR	*South Omo*	5°8'N 36°46'E
23	J2	**Bur-Mulale**	Somali	*Fafan*	8°59'N 42°55'E
23	J2	**Bur-Naf**	Somali	*Fafan*	8°55'N 42°57'E
41	G3	**Buruole**	Afar	*Kilbati (Zone 2)*	13°17'N 39°55'E
10	C3	**Butta**	SNNPR	*Kaffa*	7°7'N 36°27'E
10	E5	**Buzzani**	SNNPR	*Gamo Gofa*	6°29'N 36°59'E
39	J3	**Bwahit (Bauhit)**	Amhara	*North Gonder*	13°15'N 38°13'E

C

18	D3	**Cabba**	Oromia	*East Wellega*	8°40'N 36°41'E
41	G1	**Cabtai**	Tigray	*Eastern*	13°44'N 39°50'E
4	E5	**Calciati**	Oromia	*Borena*	4°48'N 38°4'E
36	B3	**Calloh**	Amhara	*North Wollo*	11°36'N 39°47'E
40	J2	**Cama**	Tigray	*North Western*	13°35'N 38°23'E
40	J2	**Canechi**	Amhara	*North Gonder*	13°30'N 38°20'E
28	E5	**Caniss**	Amhara	*North Shewa (R3)*	9°26'N 39°28'E
2	C2	**Canta**	SNNPR	*Bench Maji*	6°3'N 35°17'E
38	C3	**Carcarset**	Amhara	*North Gonder*	13°7'N 36°30'E
27	I1	**Carni**	Amhara	*East Gojam*	10°45'N 38°0'E
39	G5	**Carua Mariam**	Amhara	*North Gonder*	12°35'N 37°26'E
19	H3	**Cascio**	Oromia	*South West Shewa*	8°47'N 37°50'E

Page	Grid	Name	Region	Zone	Coordinates
1	J3	**Cassi**	Oromia	*West Arsi*	7°10'N 38°18'E
9	F1	**Catri Gua**	Tigray	*Western*	13°47'N 37°19'E
9	F1	**Cefa**	Tigray	*Western*	13°58'N 37°19'E
9	F3	**Celguana**	Amhara	*North Gonder*	13°0'N 37°10'E
5	F3	**Cellalaca**	Benishangul Gumuz	*Asosa*	10°1'N 34°55'E
7	I1	**Celti**	Amhara	*East Gojam*	10°46'N 37°55'E
8	D2	**Ch'alchis Terara**	Oromia	*East Wellega*	9°8'N 36°44'E
	I3	**Ch'ali Terara**	SNNPR	*South Omo*	5°45'N 36°45'E
	H4	**Challa**	SNNPR	*South Omo*	5°23'N 36°34'E
1	H1	**Chebrit Ale**	Afar	*Kilbati (Zone 2)*	13°58'N 40°8'E
6	C5	**Chedar Amba**	Amhara	*Oromia*	11°1'N 40°3'E
4	E5	**Chef**	Amhara	*East Gojam*	11°2'N 38°14'E
4	E2	**Chelala**	Amhara	*South Gonder*	12°1'N 38°8'E
9	F2	**Chelia**	Oromia	*West Shewa*	9°6'N 37°19'E
8	E1	**Chencher**	Tigray	*Western*	13°59'N 36°58'E
4	H5	**Chendam**	Amhara	*South Gonder*	12°16'N 37°48'E
3	I2	**Chera**	Oromia	*Bale*	7°27'N 40°15'E
1	J2	**Ch'erch'er**	Oromia	*West Hararge*	8°55'N 40°35'E
9	G4	**Ch'erch'er**	Amhara	*North Gonder*	12°41'N 37°27'E
2	A2	**Ch'erch'er Terara**	Oromia	*West Hararge*	9°0'N 40°48'E
8	E2	**Chesciat Abo**	Tigray	*Western*	13°27'N 37°2'E
9	I1	**Chessad Afoton**	Tigray	*North Western*	13°48'N 38°6'E
	C2	**Chiafe**	SNNPR	*Bench Maji*	6°3'N 35°22'E
0	E3	**Chidana Micrat**	Tigray	*Southern*	13°18'N 39°29'E
2	D2	**Ch'ik'e Terara**	Oromia	*Arsi*	7°30'N 39°1'E
0	E5	**Ch'ilalo Terara**	Oromia	*Arsi*	7°55'N 39°16'E
7	G1	**Ch'ok'e**	Amhara	*West Gojam*	10°45'N 37°35'E
7	H1	**Ch'ok'e Terara**	Amhara	*East Gojam*	10°43'N 37°51'E
5	H3	**Ciababa**	Benishangul Gumuz	*Metekel*	10°13'N 35°22'E
2	A3	**Ciabbi**	Oromia	*West Arsi*	7°11'N 38°26'E
1	F5	**Ciai**	Oromia	*Arsi*	8°2'N 39°31'E
9	G2	**Cialud**	Amhara	*North Gonder*	13°20'N 37°31'E
9	G3	**Ciancio Gefersa**	Oromia	*West Shewa*	8°48'N 37°24'E
9	H3	**Ciarduqua**	Amhara	*North Gonder*	13°15'N 37°39'E
8	E1	**Cimbia**	Amhara	*South Wollo*	10°52'N 39°16'E
5	G3	**Ciono**	Benishangul Gumuz	*Asosa*	9°58'N 35°8'E
6	C3	**Ciuraccia**	Oromia	*Kelem Wellega*	8°44'N 34°15'E
6	A1	**Cobat**	Tigray	*Southern*	12°17'N 39°33'E
8	C4	**Codda**	Oromia	*Ilubabor*	8°11'N 36°36'E
0	A3	**Cofla**	Oromia	*South West Shewa*	8°31'N 38°27'E
1	F2	**Cogi**	Amhara	*North Shewa (R3)*	8°52'N 39°38'E
9	F1	**Cola Siba**	Tigray	*Western*	13°39'N 37°15'E
9	G4	**Colbo**	SNNPR	*Gurage*	8°17'N 37°32'E
5	I4	**Collo**	Amhara	*South Wollo*	11°10'N 39°10'E
9	I3	**Combolata**	Oromia	*South West Shewa*	8°33'N 38°3'E
	G4	**Comeca**	Gambella	*Agnuak*	6°47'N 35°6'E
8	E2	**Conci**	Oromia	*East Wellega*	8°51'N 37°4'E
6	A3	**Corac**	Benishangul Gumuz	*Kemashi*	10°3'N 35°58'E
	C3	**Corma**	SNNPR	*Bench Maji*	5°45'N 35°21'E
4	E5	**Corosco**	Oromia	*West Wellega*	9°27'N 34°34'E
3	F3	**Cotera**	Oromia	*Bale*	7°4'N 39°42'E
4	A2	**Cubbaio**	Oromia	*Bale*	7°36'N 40°47'E
5	I2	**Culan Sancai**	Benishangul Gumuz	*Metekel*	10°35'N 35°45'E
6	C5	**Cuma**	Oromia	*East Wellega*	9°33'N 36°35'E
9	G2	**Cumbi**	Amhara	*North Shewa (R3)*	10°27'N 39°49'E
	A4	**Cuntulo**	Oromia	*Guji*	5°25'N 39°38'E
8	A5	**Cuora**	Amhara	*North Gonder*	12°25'N 36°4'E
9	F3	**Curcuana**	Amhara	*North Gonder*	13°0'N 37°10'E
9	F3	**Curcura**	Amhara	*North Gonder*	13°5'N 37°19'E
1	I4	**Dabbaho**	Afar	*Fenti (Zone 4)*	12°45'N 40°27'E
1	J5	**Dabe Terara**	Oromia	*West Hararge*	8°4'N 40°34'E
3	H4	**Dabei**	Oromia	*Bale*	6°56'N 40°14'E
7	F4	**Daben Terara**	Oromia	*Horo Guduru*	9°55'N 37°10'E
8	B3	**Dabo Gumbi**	Oromia	*Ilubabor*	8°46'N 36°17'E
	I5	**Dacai Adade**	Somali	*Afder*	5°5'N 41°30'E
1	G1	**Dagah Lebi**	Somali	*Siti*	10°59'N 42°18'E
7	G1	**Dagat**	Afar	*Awsi (Zone 1)*	12°25'N 41°7'E
39	G2	**Dagua Tafa**	Amhara	*North Gonder*	13°19'N 37°23'E
27	H2	**Daguai**	Amhara	*East Gojam*	10°38'N 37°43'E
41	F1	**Dagudit YeTerara Senselet**	Tigray	*Eastern*	13°50'N 39°34'E
30	C2	**Dahau**	Somali	*Siti*	10°32'N 41°7'E
26	C5	**Dale Terara**	Oromia	*East Wellega*	9°33'N 36°27'E
21	J1	**Daleccia**	Oromia	*West Hararge*	9°14'N 40°40'E
23	G3	**Daleti Afeta Terara**	Oromia	*East Hararge*	8°35'N 42°8'E
18	D1	**Daleti Terara**	Oromia	*East Wellega*	9°16'N 36°41'E
22	B1	**Dalo Terara**	Oromia	*West Hararge*	9°22'N 41°5'E
31	I1	**Damer**	Somali	*Siti*	10°52'N 42°51'E
15	H4	**Damerta**	Somali	*Shabelle*	6°39'N 42°35'E
2	C2	**Dami**	SNNPR	*Bench Maji*	6°6'N 35°24'E
26	A4	**Dami**	Benishangul Gumuz	*Kemashi*	9°48'N 36°5'E
13	I4	**Damole Terara**	Oromia	*Bale*	6°45'N 40°27'E
10	D3	**Damota**	SNNPR	*Konta*	7°3'N 36°39'E
11	H4	**Damota Terara**	SNNPR	*Wolayita*	6°55'N 37°47'E
20	A4	**Damu Terara**	SNNPR	*Gurage*	8°12'N 38°26'E
36	D3	**Damuta Terara**	Afar	*Awsi (Zone 1)*	11°42'N 40°11'E
21	H3	**Dana**	Oromia	*West Hararge*	8°38'N 40°8'E
31	I4	**Dananley**	Oromia	*East Hararge*	9°38'N 42°51'E
33	F4	**Dangur**	Benishangul Gumuz	*Metekel*	11°22'N 35°57'E
21	H3	**Danse**	Oromia	*Arsi*	8°30'N 40°9'E
28	A5	**Danse**	Oromia	*West Shewa*	9°25'N 38°20'E
18	A2	**Dapo Gumbi**	Oromia	*Ilubabor*	8°54'N 36°5'E
39	F2	**Dara**	Tigray	*Western*	13°19'N 37°15'E
5	J3	**Darar**	Oromia	*Guji*	5°36'N 39°20'E
26	C1	**Darim**	Amhara	*Awi/Agew*	10°44'N 36°35'E
31	H4	**Darimi YeTerara Senselet**	Oromia	*East Hararge*	9°42'N 42°29'E
22	A4	**Daro Lebu YeTerara Senselet**	Oromia	*West Hararge*	8°7'N 40°44'E
21	J5	**Daroti Terara**	Oromia	*West Hararge*	7°58'N 40°35'E
4	D4	**Darrito**	Oromia	*Borena*	5°11'N 38°2'E
4	B1	**Darsi**	SNNPR	*Gamo Gofa*	6°17'N 37°33'E
18	A4	**Darumu**	Oromia	*Ilubabor*	8°21'N 36°7'E
34	D1	**Dauc**	Amhara	*North Gonder*	12°15'N 37°57'E
37	G2	**Daurre**	Afar	*Awsi (Zone 1)*	11°57'N 41°1'E
39	I3	**Debark'**	Amhara	*North Gonder*	13°11'N 37°57'E
20	A3	**Debel Terara**	Oromia	*South West Shewa*	8°48'N 38°28'E
27	I2	**Debet**	Amhara	*East Gojam*	10°29'N 38°1'E
20	E3	**Debocogio**	Amhara	*North Shewa (R3)*	8°45'N 39°18'E
34	D3	**Debra Chente**	Amhara	*South Gonder*	11°31'N 37°55'E
40	B1	**Debra Ghennet**	Tigray	*Central*	13°54'N 38°38'E
39	I2	**Debra Tiun**	Amhara	*North Gonder*	13°22'N 38°4'E
40	C1	**Debre Asa YeTerara Senselet**	Tigray	*Central*	13°44'N 38°58'E
40	E1	**Debre Ma'ir YeTerara Ch'af**	Tigray	*Eastern*	13°53'N 39°23'E
40	E1	**Debre Mar Terara**	Tigray	*Eastern*	13°52'N 39°24'E
35	F2	**Debre Sina**	Amhara	*South Gonder*	11°58'N 38°30'E
27	I5	**Degat'ina Amba**	Oromia	*West Shewa*	9°32'N 37°58'E
26	E3	**Degga**	Oromia	*Horo Guduru*	10°8'N 37°4'E
16	E2	**Deka Jeldesa Terara**	Oromia	*Kelem Wellega*	8°55'N 34°40'E
26	A5	**Delgagala**	Benishangul Gumuz	*Kemashi*	9°27'N 36°6'E
11	G5	**Delle**	SNNPR	*Gamo Gofa*	6°28'N 37°26'E
4	D2	**Delo**	SNNPR	*Segen Peoples'*	5°49'N 37°50'E
37	I4	**Dema'ali Terara**	Afar	*Awsi (Zone 1)*	11°17'N 41°38'E
36	C3	**Dembel**	Amhara	*South Wollo*	11°29'N 40°0'E
22	B1	**Dembi Terara**	Oromia	*West Hararge*	9°21'N 41°4'E
39	F3	**Dendelcau**	Tigray	*Western*	13°17'N 37°13'E
19	H2	**Dendi**	Oromia	*West Shewa*	8°49'N 37°51'E
19	I2	**Dendi Terara**	Oromia	*West Shewa*	8°49'N 38°2'E
31	F5	**Dera Terara**	Dire Dawa	*Dire Dawa*	9°34'N 41°54'E
29	I4	**Derabale Terara**	Afar	*Gabi (Zone 3)*	9°45'N 40°17'E
31	F4	**Dere Meley YeTerara Senselet**	Somali	*Siti*	9°41'N 42°2'E
35	F5	**Deri**	Amhara	*East Gojam*	10°55'N 38°21'E
19	G2	**Derico**	Oromia	*West Shewa*	8°56'N 37°36'E
5	I5	**Derredima**	Oromia	*Borena*	4°53'N 39°11'E
31	H1	**Deryo**	Somali	*Siti*	10°52'N 42°38'E
16	E3	**Dewi Terara**	Oromia	*Kelem Wellega*	8°41'N 34°42'E
31	H5	**Dibriuein**	Oromia	*East Hararge*	9°35'N 42°33'E

18	C2	**Digga**	Oromia	*East Wellega*	9°2'N 36°29'E
36	C1	**Dighim**	Afar	*Fenti (Zone 4)*	12°14'N 40°10'E
40	D2	**Dilarye Terara**	Tigray	*Central*	13°36'N 39°1'E
25	G3	**Dildessa**	Benishangul Gumuz	*Asosa*	10°15'N 35°8'E
22	B3	**Dimbiba YeTerara Senselet**	Oromia	*West Hararge*	8°40'N 41°2'E
40	D1	**Dira Amba Terara**	Tigray	*Central*	13°39'N 39°2'E
18	D5	**Dire YeTerara Senselet**	Oromia	*East Wellega*	9°18'N 36°48'E
4	B1	**Dita**	SNNPR	*Gamo Gofa*	6°14'N 37°28'E
31	J4	**Djebi Abeyou**	Somali	*Fafan*	9°51'N 42°59'E
22	D1	**Djemi**	Oromia	*East Hararge*	9°21'N 41°34'E
28	D3	**Doba**	Amhara	*North Shewa (R3)*	10°3'N 39°5'E
31	G3	**Doba Terara**	Somali	*Siti*	9°59'N 42°16'E
37	I2	**Dobi YeTerara Senselet**	Afar	*Awsi (Zone 1)*	11°58'N 41°32'E
12	D4	**Doda**	Oromia	*West Arsi*	6°54'N 39°14'E
21	J5	**Dodota Terara**	Oromia	*West Hararge*	7°59'N 40°33'E
21	H5	**Dofen Terara**	Afar	*Gabi (Zone 3)*	9°20'N 40°8'E
11	F5	**Doghe**	SNNPR	*Gamo Gofa*	6°36'N 37°23'E
32	B5	**Dong**	Benishangul Gumuz	*Kemashi*	11°0'N 35°3'E
19	F2	**Dongoro YeTerara Senselet**	Oromia	*West Shewa*	9°8'N 37°15'E
39	G4	**Dredebbe**	Amhara	*North Gonder*	12°42'N 37°35'E
9	G4	**Duco**	SNNPR	*Bench Maji*	6°54'N 35°10'E
41	F5	**Duffat**	Tigray	*Southern*	12°30'N 39°30'E
11	I4	**Duguna**	SNNPR	*Wolayita*	6°57'N 38°3'E
24	D2	**Dul**	Benishangul Gumuz	*Asosa*	10°29'N 34°24'E
31	H4	**Duleti Terara**	Somali	*Siti*	9°44'N 42°28'E
10	D3	**Dulla**	SNNPR	*Kaffa*	7°9'N 36°38'E
25	H2	**Duma**	Benishangul Gumuz	*Metekel*	10°20'N 35°22'E
34	D1	**Dumi**	Amhara	*South Gonder*	12°14'N 37°50'E
22	C1	**Dunk'uk'u Terara**	Oromia	*West Hararge*	9°10'N 41°9'E
31	G1	**Durdur**	Somali	*Siti*	10°52'N 42°16'E
4	J3	**Duro**	SNNPR	*Segen Peoples'*	5°31'N 37°12'E
12	C3	**Duro Terara**	Oromia	*West Arsi*	7°14'N 38°49'E

E

39	G2	**Ecimcia**	Amhara	*North Gonder*	13°19'N 37°25'E
4	C3	**Edulei**	SNNPR	*Segen Peoples'*	5°39'N 37°40'E
19	F4	**Egan**	Oromia	*Jimma*	8°25'N 37°16'E
20	E1	**Egdu**	Amhara	*North Shewa (R3)*	9°17'N 39°18'E
41	F2	**Eghir Eriva**	Tigray	*Southern*	13°28'N 39°40'E
31	G1	**Egle**	Somali	*Siti*	10°46'N 42°18'E
19	J4	**Egu**	SNNPR	*Gurage*	8°22'N 38°11'E
25	G1	**El Danab**	Benishangul Gumuz	*Asosa*	10°41'N 35°4'E
14	D4	**El Uak**	Oromia	*Bale*	6°51'N 41°35'E
41	G3	**Ela Callo**	Afar	*Kilbati (Zone 2)*	13°6'N 40°0'E
31	H3	**Eles-Terara**	Somali	*Siti*	10°12'N 42°36'E
33	G3	**Elfing Bar**	Amhara	*Awi/Agew*	11°47'N 36°15'E
6	A5	**Eluddoro**	Oromia	*Guji*	4°48'N 39°38'E
28	A2	**Embeleti**	Amhara	*East Gojam*	10°37'N 38°25'E
39	J3	**Encetcab**	Amhara	*North Gonder*	13°6'N 38°9'E
38	D4	**Encheddiba**	Amhara	*North Gonder*	12°37'N 36°37'E
25	I3	**Enemo**	Benishangul Gumuz	*Kemashi*	9°57'N 35°46'E
31	I1	**Enlon-Godon**	Somali	*Siti*	10°45'N 42°41'E
31	H1	**Erene**	Somali	*Siti*	10°56'N 42°36'E
20	C2	**Erer Terara**	Oromia	*East Shewa*	8°54'N 38°58'E
39	H4	**Ernambe**	Amhara	*North Gonder*	12°55'N 37°38'E
12	E1	**Erosa**	Oromia	*Arsi*	7°51'N 39°23'E
36	D1	**Errerei**	Afar	*Fenti (Zone 4)*	12°9'N 40°19'E
24	E3	**Eruoldi**	Benishangul Gumuz	*Asosa*	10°13'N 34°37'E
39	G1	**Eto**	Tigray	*Western*	13°40'N 37°23'E
35	I2	**Ezeran**	Amhara	*North Wollo*	12°1'N 39°11'E
34	D2	**Ezub**	Amhara	*South Gonder*	12°6'N 37°48'E
25	G2	**Fado**	Benishangul Gumuz	*Kemashi*	10°29'N 35°7'E

F

32	B5	**Fadokha**	Benishangul Gumuz	*Kemashi*	10°57'N 34°58'E
25	G3	**Fagaddo**	Benishangul Gumuz	*Asosa*	10°11'N 35°9'E
21	F4	**Fagogi**	Oromia	*Arsi*	8°18'N 39°39'E
19	I2	**Faicha**	Oromia	*West Shewa*	8°56'N 38°0'E
23	F2	**Falti**	Oromia	*East Hararge*	9°2'N 41°54'E
24	D3	**Famoghe**	Benishangul Gumuz	*Asosa*	10°15'N 34°24'E
25	F3	**Farboghi**	Benishangul Gumuz	*Asosa*	10°9'N 34°47'E
25	G3	**Fari**	Benishangul Gumuz	*Asosa*	10°15'N 35°7'E
30	A5	**Farsis**	Afar	*Gabi (Zone 3)*	9°27'N 40°41'E
3	I3	**Fascalli**	SNNPR	*South Omo*	5°30'N 36°46'E
22	E1	**Fedni Terara**	Oromia	*East Hararge*	9°9'N 41°48'E
19	F4	**Fego**	Oromia	*Jimma*	8°9'N 37°12'E
40	E3	**Felege Hafo Terara**	Tigray	*Southern*	13°16'N 39°21'E
21	G2	**Fentale Isate Gemora**	Oromia	*East Shewa*	8°59'N 39°56'E
39	I3	**Feraca**	Amhara	*North Gonder*	13°12'N 37°54'E
31	I2	**Feras**	Somali	*Siti*	10°27'N 42°48'E
23	G4	**Fik'**	Somali	*Nogob*	8°7'N 42°18'E
12	E2	**Filfo Terara**	Oromia	*Arsi*	7°22'N 39°21'E
40	E3	**Fingalat**	Tigray	*Southern*	13°9'N 39°23'E
20	B1	**Fota Terara**	Oromia	*West Shewa*	9°13'N 38°30'E
33	I5	**Fudi**	Amhara	*Awi/Agew*	10°55'N 36°47'E
17	I5	**Fugu**	Oromia	*Ilubabor*	8°3'N 35°33'E
33	H1	**Fuldana**	Amhara	*North Gonder*	12°10'N 36°28'E
5	I4	**Fulu**	Oromia	*Guji*	5°8'N 39°12'E
31	H5	**Funyan Ch'ali Terara**	Oromia	*East Hararge*	9°24'N 42°30'E
31	H5	**Funyan Jufu Terara**	Oromia	*East Hararge*	9°24'N 42°26'E
36	C3	**Fura**	Afar	*Awsi (Zone 1)*	11°44'N 40°4'E
22	B4	**Furdo Terara**	Oromia	*West Hararge*	8°11'N 40°55'E

G

38	D2	**Gababu**	Tigray	*Western*	13°27'N 36°45'E
18	E4	**Gabana**	Oromia	*Jimma*	8°8'N 36°53'E
37	H5	**Gabillema**	Afar	*Awsi (Zone 1)*	11°5'N 41°16'E
33	H2	**Gabina**	Amhara	*Awi/Agew*	11°58'N 36°26'E
41	H1	**Gabuli**	Afar	*Kilbati (Zone 2)*	13°40'N 40°16'E
4	B2	**Gacio**	SNNPR	*Gamo Gofa*	6°9'N 37°32'E
37	G1	**Gaddaclo**	Afar	*Awsi (Zone 1)*	12°13'N 40°59'E
2	C2	**Gafala**	SNNPR	*Bench Maji*	6°6'N 35°22'E
39	G4	**Gai**	Amhara	*North Gonder*	12°42'N 37°29'E
8	E4	**Gak**	Gambella	*Agnuak*	6°51'N 34°38'E
37	I2	**Gala**	Afar	*Awsi (Zone 1)*	11°51'N 41°41'E
21	F2	**Gala Terara**	Amhara	*North Shewa (R3)*	8°50'N 39°32'E
9	I4	**Galajiba**	SNNPR	*Bench Maji*	6°45'N 35°39'E
29	J4	**Galalu**	Afar	*Gabi (Zone 3)*	9°47'N 40°38'E
12	E2	**Galama**	Oromia	*Arsi*	7°39'N 39°22'E
37	G4	**Galatu**	Afar	*Awsi (Zone 1)*	11°17'N 41°6'E
38	C5	**Galea**	Amhara	*North Gonder*	12°25'N 36°26'E
36	D1	**Galicoma**	Afar	*Fenti (Zone 4)*	12°12'N 40°15'E
36	D1	**Gamabolu**	Afar	*Fenti (Zone 4)*	12°21'N 40°15'E
11	H2	**Gambata**	SNNPR	*Kembata Tembaro*	7°20'N 37°50'E
36	C2	**Gambato**	Afar	*Fenti (Zone 4)*	11°51'N 40°3'E
4	E2	**Gamso**	SNNPR	*Gedio*	6°3'N 38°8'E
35	H2	**Gana**	Amhara	*North Wollo*	12°1'N 38°52'E
38	C4	**Gandeba**	Amhara	*North Gonder*	12°41'N 36°23'E
33	I2	**Gandi**	Amhara	*North Gonder*	11°58'N 36°48'E
30	D4	**Gara Adi Terara**	Somali	*Siti*	9°46'N 41°34'E
12	C1	**Gara Alutu**	Oromia	*Arsi*	7°46'N 38°47'E
31	H5	**Gara Amola**	Oromia	*East Hararge*	9°33'N 42°30'E
14	D2	**Gara Colba**	Oromia	*Bale*	7°28'N 41°22'E
20	B2	**Gara Furi Terara**	Oromia	*South West Shewa*	8°53'N 38°41'E
31	I4	**Gara Garri**	Somali	*Fafan*	9°43'N 42°50'E
31	I5	**Gara Goreis**	Oromia	*East Hararge*	9°28'N 42°38'E
28	B4	**Gara Guda Terara**	Oromia	*North Shewa (R4)*	9°47'N 38°38'E
22	B2	**Gara K'ufa**	Oromia	*West Hararge*	9°8'N 41°5'E
22	E1	**Gara Muleta**	Oromia	*East Hararge*	9°15'N 41°44'E
25	I4	**Gara Nasa**	Benishangul Gumuz	*Kemashi*	9°42'N 35°33'E
12	E2	**Gara Sero**	Oromia	*Arsi*	7°36'N 39°22'E
23	G5	**Gara-Fik-Uen**	Somali	*Nogob*	8°4'N 42°20'E
41	F3	**Garagiam**	Tigray	*Southern*	13°6'N 39°37'E
26	D4	**Garamate**	Oromia	*East Wellega*	9°44'N 36°39'E
12	B4	**Garamba**	SNNPR	*Sidama*	6°44'N 38°39'E
33	H2	**Garasgh**	Amhara	*North Gonder*	12°0'N 36°39'E

38	C2	**Garasitu**	Amhara	*North Gonder*	13°20'N 36°28'E
41	I2	**Garbena**	Afar	*Kilbati (Zone 2)*	13°27'N 40°24'E
36	C5	**Garfa**	Amhara	*Oromia*	11°5'N 40°1'E
11	F2	**Garo**	Oromia	*Jimma*	7°25'N 37°17'E
36	B1	**Garra Aleucia**	Amhara	*North Wollo*	12°9'N 39°41'E
18	D5	**Garugge**	Oromia	*Jimma*	8°2'N 36°52'E
41	I4	**Garule**	Afar	*Kilbati (Zone 2)*	12°55'N 40°17'E
35	J2	**Gaso**	Amhara	*North Wollo*	11°50'N 39°21'E
10	B1	**Gaua**	Oromia	*Jimma*	7°39'N 36°8'E
35	H1	**Gavsigivla**	Amhara	*Wag Himra*	12°21'N 39°4'E
33	H1	**Gaz Gimb**	Amhara	*North Gonder*	12°9'N 36°35'E
24	D3	**Gaza**	Benishangul Gumuz	*Asosa*	10°5'N 34°22'E
40	B1	**Gebaco**	Tigray	*Central*	13°49'N 38°38'E
23	I4	**Gebel Goreis**	Oromia	*East Hararge*	8°26'N 42°39'E
38	E2	**Gebre Harya Terara**	Tigray	*Western*	13°26'N 37°2'E
39	F2	**Gebre Harya Terara**	Tigray	*Western*	13°20'N 37°12'E
11	I4	**Gedano**	SNNPR	*Sidama*	6°38'N 38°8'E
37	G1	**Ged'elu Terara**	Afar	*Awsi (Zone 1)*	12°24'N 41°4'E
22	B5	**Gedera Terara**	Oromia	*West Hararge*	8°4'N 40°55'E
26	B5	**Gello**	Benishangul Gumuz	*Kemashi*	9°33'N 36°12'E
11	F4	**Genei**	SNNPR	*Dawro*	6°45'N 37°11'E
36	D3	**Gengoyta Terara**	Afar	*Awsi (Zone 1)*	11°41'N 40°21'E
4	B2	**Genta Terara**	SNNPR	*Gamo Gofa*	6°4'N 37°27'E
40	D5	**Gerbaco**	Amhara	*Wag Himra*	12°26'N 39°5'E
20	C3	**Gerbicha Terara**	Oromia	*East Shewa*	8°41'N 38°57'E
40	C3	**Gerum**	Tigray	*Central*	13°16'N 38°57'E
20	A4	**Gesh Megal Terara**	Oromia	*South West Shewa*	8°24'N 38°24'E
9	J1	**Gesha**	SNNPR	*Kaffa*	7°44'N 35°52'E
23	F2	**Geyle Terara**	Oromia	*East Hararge*	9°3'N 41°52'E
39	F1	**Geza Agel**	Tigray	*Western*	13°56'N 37°7'E
38	D2	**Ghecciau**	Amhara	*North Gonder*	13°22'N 36°49'E
36	B1	**Ghedaro**	Afar	*Fenti (Zone 4)*	12°12'N 39°49'E
40	J4	**Ghedeon**	Amhara	*North Gonder*	12°38'N 38°23'E
39	G4	**Ghedi**	Amhara	*North Gonder*	12°39'N 37°31'E
14	D4	**Ghelbi**	Oromia	*Bale*	6°47'N 41°27'E
17	J5	**Ghermana**	Oromia	*Jimma*	7°54'N 35°56'E
4	E2	**Ghibigia**	Oromia	*Borena*	6°2'N 38°5'E
37	I3	**Ghifa**	Afar	*Awsi (Zone 1)*	11°37'N 41°40'E
41	H4	**Ghimmeri**	Afar	*Kilbati (Zone 2)*	12°56'N 40°12'E
14	E5	**Giabis**	Somali	*Afder*	6°23'N 41°36'E
40	D5	**Giala**	Amhara	*Wag Himra*	12°24'N 39°4'E
11	G1	**Giangero**	SNNPR	*Yem*	7°45'N 37°28'E
39	H3	**Gianora**	Amhara	*North Gonder*	13°2'N 37°37'E
18	E1	**Giaue**	Oromia	*West Shewa*	9°9'N 37°3'E
26	B4	**Gibir**	Benishangul Gumuz	*Kemashi*	9°44'N 36°8'E
22	E2	**Gicha Terara**	Oromia	*East Hararge*	8°51'N 41°51'E
26	E3	**Gidda**	Oromia	*East Wellega*	10°0'N 36°52'E
33	G2	**Gilfu**	Amhara	*North Gonder*	12°6'N 36°12'E
34	C3	**Gina**	Amhara	*South Gonder*	11°35'N 37°41'E
15	H3	**Giogo**	Somali	*Nogob*	7°10'N 42°35'E
13	H4	**Goba**	Oromia	*Bale*	6°55'N 40°10'E
31	G3	**Goba-Dzher**	Somali	*Siti*	10°1'N 42°12'E
31	H5	**Gobbi Hamer**	Oromia	*East Hararge*	9°24'N 42°34'E
30	E2	**Gobe'a Terara**	Somali	*Siti*	10°29'N 41°44'E
3	I2	**Goda**	SNNPR	*South Omo*	5°50'N 36°54'E
10	A3	**Goda**	SNNPR	*Kaffa*	7°10'N 36°5'E
31	I1	**Godire**	Somali	*Siti*	10°56'N 42°50'E
13	H2	**Gogiabe**	Oromia	*Arsi*	7°30'N 40°5'E
15	G4	**Goja Terara**	Somali	*Shabelle*	6°46'N 42°9'E
12	A2	**Golgio**	SNNPR	*Alaba*	7°32'N 38°21'E
38	E2	**Golgoda**	Amhara	*North Gonder*	13°19'N 37°1'E
26	D4	**Gomboch'Terara**	Oromia	*East Wellega*	9°45'N 36°43'E
39	G3	**Gombolo**	Amhara	*North Gonder*	13°11'N 37°22'E
36	J5	**Gomoyta YeTerara Senselet**	Afar	*Fenti (Zone 4)*	12°18'N 40°32'E
26	C4	**Gondor Terara**	Oromia	*East Wellega*	9°53'N 36°33'E
33	H5	**Gongi**	Amhara	*Awi/Agew*	10°58'N 36°38'E
18	C2	**Gonjo**	Oromia	*East Wellega*	9°2'N 36°23'E
9	H4	**Gorei**	SNNPR	*Bench Maji*	6°38'N 35°27'E
20	C1	**Gorfo Terara**	Oromia	*North Shewa (R4)*	9°15'N 38°52'E
10	A1	**Gorgota**	Oromia	*Jimma*	7°45'N 36°6'E
26	E5	**Goroch'an**	Oromia	*Horo Guduru*	9°26'N 37°4'E
22	E3	**Gorogarbi**	Oromia	*East Hararge*	8°41'N 41°48'E
23	F3	**Gorru**	Oromia	*East Hararge*	8°33'N 42°4'E
12	C4	**Gorte Terara**	Oromia	*West Arsi*	6°50'N 38°50'E
26	E4	**Gorte Terara**	Oromia	*Horo Guduru*	9°38'N 36°56'E
18	B1	**Gotto Muri**	Benishangul Gumuz	*Kemashi*	9°12'N 36°10'E
40	E3	**Gowdi Hayrat Terara**	Tigray	*Southern*	13°7'N 39°26'E
10	B3	**Guba**	SNNPR	*Kaffa*	7°11'N 36°10'E
36	A4	**Guba YeTerara Senselet**	Amhara	*South Wollo*	11°12'N 39°37'E
4	D4	**Gubba**	Oromia	*Borena*	5°25'N 37°58'E
26	C2	**Gubbi**	Benishangul Gumuz	*Kemashi*	10°19'N 36°36'E
25	I2	**Guciar**	Benishangul Gumuz	*Metekel*	10°29'N 35°38'E
20	E2	**Guga**	Amhara	*North Shewa (R3)*	9°3'N 39°26'E
4	B1	**Guge**	SNNPR	*Gamo Gofa*	6°12'N 37°30'E
3	G3	**Guista**	SNNPR	*South Omo*	5°38'N 36°28'E
20	C3	**Guji Terara**	Oromia	*East Shewa*	8°47'N 38°50'E
24	E2	**Gule**	Benishangul Gumuz	*Asosa*	10°30'N 34°31'E
22	B1	**Gulube Terara**	Oromia	*West Hararge*	9°19'N 40°55'E
26	C2	**Gum**	Amhara	*Awi/Agew*	10°36'N 36°35'E
17	G4	**Guma**	Oromia	*Ilubabor*	8°17'N 35°2'E
17	G2	**Guma Guda YeTerara Ch'af**	Oromia	*Kelem Wellega*	9°1'N 35°7'E
39	G1	**Gumai Daro**	Tigray	*Western*	13°45'N 37°28'E
31	H1	**Gumar**	Somali	*Siti*	10°56'N 42°27'E
41	F4	**Gumasir**	Tigray	*Southern*	12°56'N 39°44'E
21	H2	**Gumbi**	Oromia	*West Hararge*	8°58'N 40°13'E
38	D4	**Gumenta**	Amhara	*North Gonder*	12°54'N 36°38'E
36	D1	**Gummeli**	Afar	*Fenti (Zone 4)*	12°12'N 40°24'E
34	E3	**Guna Terara**	Amhara	*South Gonder*	11°43'N 38°14'E
29	F4	**Gundi**	Amhara	*North Shewa (R3)*	9°46'N 39°43'E
4	B5	**Gundile**	Oromia	*Borena*	4°54'N 37°24'E
21	H4	**Gura Terara**	Oromia	*Arsi*	8°8'N 40°13'E
36	D3	**Gura'ale Terara**	Afar	*Awsi (Zone 1)*	11°40'N 40°25'E
17	H4	**Guraccio**	Oromia	*Ilubabor*	8°26'N 35°20'E
9	F4	**Guraferda YeTerara Senselet**	SNNPR	*Bench Maji*	6°55'N 34°59'E
20	A4	**Gurage**	Oromia	*South West Shewa*	8°24'N 38°24'E
9	I4	**Gurboji**	SNNPR	*Bench Maji*	6°41'N 35°43'E
9	J2	**Gurguri**	SNNPR	*Kaffa*	7°21'N 35°54'E
9	H3	**Gurma**	SNNPR	*Bench Maji*	6°58'N 35°18'E
4	A3	**Guroza Terara**	SNNPR	*Segen Peoples'*	5°35'N 37°17'E
18	E1	**Gushis YeTerara Ch'af**	Oromia	*East Wellega*	9°18'N 36°57'E
19	H3	**Guto**	Oromia	*South West Shewa*	8°43'N 37°45'E
26	C2	**Gutr**	Amhara	*West Gojam*	10°30'N 36°35'E
29	G2	**Gwasa Terara**	Amhara	*North Shewa (R3)*	10°25'N 39°53'E

H

13	I4	**Habrona Terara**	Oromia	*Bale*	6°41'N 40°27'E
7	I2	**Hada Kumbi**	Somali	*Afder*	5°58'N 41°39'E
41	F3	**Hadele Terara**	Tigray	*Southern*	13°8'N 39°41'E
41	F4	**Haia Olel**	Tigray	*Southern*	12°43'N 39°44'E
41	J4	**Halaite**	Afar	*Fenti (Zone 4)*	12°37'N 40°34'E
37	G4	**Hallu**	Afar	*Awsi (Zone 1)*	11°23'N 41°9'E
31	G5	**Hama YeTerara Senselet**	Oromia	*East Hararge*	9°32'N 42°10'E
36	C3	**Hamarko Terara**	Amhara	*South Wollo*	11°36'N 39°58'E
26	D5	**Hambule Terara**	Oromia	*East Wellega*	9°28'N 36°50'E
19	I3	**Hamdo**	Oromia	*West Shewa*	8°48'N 37°54'E
36	A5	**Hara YeTerara Senselet**	Amhara	*South Wollo*	11°4'N 39°36'E
37	F1	**Hararu Terara**	Afar	*Awsi (Zone 1)*	12°10'N 40°49'E
21	F3	**Harbona Terara**	Oromia	*East Shewa*	8°42'N 39°35'E
39	G3	**Haria**	Amhara	*North Gonder*	13°3'N 37°31'E
31	G2	**Harr**	Somali	*Siti*	10°28'N 42°18'E
39	I2	**Haweza Terara**	Amhara	*North Gonder*	13°25'N 38°4'E
40	B2	**Hay**	Amhara	*North Gonder*	13°21'N 38°31'E
37	I4	**Haytankoma Terara**	Afar	*Awsi (Zone 1)*	11°21'N 41°35'E
39	F1	**Hemberti**	Tigray	*Western*	13°55'N 37°9'E
23	H1	**Herreder**	Oromia	*East Hararge*	9°10'N 42°23'E
18	D4	**Hesa**	Oromia	*Ilubabor*	8°14'N 36°41'E

36	A1	**Hobat**	Amhara	*North Wollo*	12°12'N 39°34'E
37	F1	**Hodale Terara**	Afar	*Awsi (Zone 1)*	12°21'N 40°44'E
7	J2	**Hogdu**	Somali	*Afder*	6°6'N 41°51'E
19	F4	**Hola**	Oromia	*Jimma*	8°16'N 37°15'E
12	B4	**Holo Terara**	SNNPR	*Sidama*	6°44'N 38°34'E
12	B5	**Honku**	SNNPR	*Sidama*	6°27'N 38°33'E
13	G3	**Hora YeTerara Senselet**	Oromia	*Bale*	7°15'N 39°45'E
31	H1	**Huglo**	Somali	*Siti*	10°44'N 42°23'E
25	J2	**Huladura Terara**	Benishangul Gumuz	*Metekel*	10°18'N 35°49'E
3	H5	**Humu YeTerara Senselet**	SNNPR	*South Omo*	5°0'N 36°40'E
36	D4	**Hurmat Terara**	Afar	*Awsi (Zone 1)*	11°24'N 40°13'E

I

25	H3	**Iagada**	Benishangul Gumuz	*Metekel*	10°17'N 35°22'E
37	J3	**Iaghar**	Afar	*Awsi (Zone 1)*	11°47'N 41°44'E
25	H2	**Iagona**	Benishangul Gumuz	*Metekel*	10°33'N 35°21'E
26	C3	**Ibantu**	Oromia	*East Wellega*	10°10'N 36°25'E
29	F1	**Ieciuol**	Amhara	*South Wollo*	10°50'N 39°30'E
12	D4	**Ilale Terara**	Oromia	*West Arsi*	6°48'N 39°9'E
21	G3	**Ilko Terara**	Oromia	*Arsi*	8°32'N 39°45'E
41	I3	**Imba**	Afar	*Kilbati (Zone 2)*	13°16'N 40°27'E
40	D2	**Imba Shihak'Terara**	Tigray	*Southern*	13°19'N 39°13'E
40	E3	**Imba Ts'emba Terara**	Tigray	*Southern*	13°2'N 39°18'E
40	E2	**Imba Zerfent Korebta**	Tigray	*Central*	13°38'N 39°16'E
41	F1	**Inda Maryam**	Tigray	*Eastern*	13°55'N 39°35'E
36	B5	**Indode Terara**	Amhara	*South Wollo*	10°60'N 39°54'E
27	H1	**Ingia**	Amhara	*East Gojam*	10°42'N 37°42'E
31	G5	**Ink'uftu YeTerara Senselet**	Oromia	*East Hararge*	9°33'N 42°14'E
20	C2	**Int'ot'o Terara**	Addis Ababa	*Region 14*	9°4'N 38°48'E
35	I5	**Ioll**	Amhara	*South Wollo*	11°1'N 39°6'E
25	H3	**Ionghi**	Benishangul Gumuz	*Kemashi*	10°1'N 35°21'E
35	I1	**Israel Amba**	Amhara	*Wag Himra*	12°16'N 39°6'E

J

2	C1	**Jaba**	SNNPR	*Bench Maji*	6°15'N 35°14'E
38	B2	**Jabal Abu Takia**	Amhara	*North Gonder*	13°18'N 36°18'E
25	A5	**Jabal Belangashe**	Benishangul Gumuz	*Asosa*	10°48'N 34°52'E
32	B5	**Jabal Belesagado**	Benishangul Gumuz	*Asosa*	10°52'N 34°57'E
24	D1	**Jabal Gamendu**	Benishangul Gumuz	*Asosa*	10°41'N 34°24'E
24	E1	**Jabal Kashangaro**	Benishangul Gumuz	*Asosa*	10°54'N 34°36'E
24	C5	**Jabal Manga**	Benishangul Gumuz	*Asosa*	9°33'N 34°6'E
38	B4	**Jabal Maqbarah**	Amhara	*North Gonder*	12°41'N 36°11'E
2	B3	**Jabal Tomadur**	SNNPR	*Bench Maji*	5°38'N 35°8'E
24	C4	**Jabal Zazing**	Benishangul Gumuz	*Asosa*	9°35'N 34°10'E
33	F2	**Jagaighi Tarara**	Amhara	*Awi/Agew*	11°52'N 36°8'E
33	H2	**Jamifala**	Amhara	*North Gonder*	12°7'N 36°26'E
24	D4	**Jebel Bange**	Benishangul Gumuz	*Asosa*	9°43'N 34°18'E
38	B4	**Jebel Dalasa**	Amhara	*North Gonder*	12°54'N 36°11'E
24	C5	**Jebel Dubu**	Benishangul Gumuz	*Asosa*	9°25'N 34°7'E
24	D4	**Jebel Gogot**	Benishangul Gumuz	*Asosa*	9°55'N 34°15'E
24	D5	**Jebel Kasr**	Benishangul Gumuz	*Asosa*	9°25'N 34°16'E
16	D1	**Jebel Shomala**	Benishangul Gumuz	*Asosa*	9°15'N 34°26'E
33	F2	**Jebel Umm Zenegia**	Amhara	*North Gonder*	12°6'N 35°59'E
33	F1	**Jebel Umm Zenidi**	Amhara	*North Gonder*	12°15'N 35°57'E
26	A4	**Jeldesa**	Benishangul Gumuz	*Kemashi*	9°39'N 36°1'E
26	D4	**Jeldu Terara**	Oromia	*Horo Guduru*	9°39'N 36°51'E
22	B2	**Jelo Terara**	Oromia	*West Hararge*	9°0'N 40°51'E
31	G5	**Jewer YeTerara Senselet**	Oromia	*East Hararge*	9°24'N 42°11'E
19	G3	**Jibat**	Oromia	*West Shewa*	8°47'N 37°28'E
39	H4	**Jibo Terara**	Amhara	*North Gonder*	12°46'N 37°41'E
18	C2	**Jimata**	Oromia	*East Wellega*	9°7'N 36°23'E
19	I2	**Jirma**	Oromia	*West Shewa*	9°5'N 38°0'E

K

9	I3	**Kaka Terara**	SNNPR	*Bench Maji*	6°58'N 35°39'E
7	J3	**Kalech**	Somali	*Afder*	5°40'N 41°50'E
9	G4	**Kamtis**	SNNPR	*Bench Maji*	6°47'N 35°7'E
29	G2	**Kebroy Maryam Terara**	Amhara	*North Shewa (R3)*	10°27'N 39°54'E
12	D2	**K'ech'a Terara**	Oromia	*Arsi*	7°22'N 39°9'E
20	A4	**K'ech'a Terara**	Oromia	*South West Shewa*	8°18'N 38°23'E
18	D1	**Kedo YeTerara Ch'af**	Oromia	*East Wellega*	9°12'N 36°42'E
21	J2	**Kentari Terara**	Oromia	*West Hararge*	8°54'N 40°35'E
20	C2	**Kereyu Terara**	Oromia	*East Shewa*	8°53'N 38°58'E
40	D1	**K'ernale Terara**	Tigray	*Central*	13°42'N 39°4'E
31	G4	**Kerora YeTerara Senselet**	Oromia	*East Hararge*	9°39'N 42°15'E
31	G5	**K'ilisa Terara**	Oromia	*East Hararge*	9°28'N 42°13'E
39	J3	**Kidis Yared**	Amhara	*North Gonder*	13°18'N 38°20'E
39	G3	**Kirakir**	Amhara	*North Gonder*	13°15'N 37°25'E
14	E2	**Koloto**	Oromia	*Bale*	7°27'N 41°49'E
26	C5	**Koma Terara**	Benishangul Gumuz	*Kemashi*	9°29'N 36°25'E
18	C2	**Komto Terara**	Oromia	*East Wellega*	9°8'N 36°37'E
21	F2	**Kora**	Amhara	*North Shewa (R3)*	8°50'N 39°39'E
37	J5	**Korbelli**	Somali	*Siti*	10°56'N 41°56'E
12	D4	**Korduro Terara**	Oromia	*West Arsi*	6°50'N 39°9'E
13	I2	**K'ori YeTerara Ch'af**	Oromia	*Bale*	7°31'N 40°17'E
12	D5	**K'ork'ite YeTerara Senselet**	SNNPR	*Sidama*	6°25'N 39°0'E
40	D1	**Koro Terara**	Tigray	*Central*	13°42'N 39°5'E
35	J3	**Koso Amba**	Amhara	*North Wollo*	11°46'N 39°26'E
11	F4	**Koysha Terara**	SNNPR	*Dawro*	6°50'N 37°15'E
12	D2	**K'ubsa Terara**	Oromia	*Arsi*	7°25'N 38°59'E
9	G4	**Kuda**	SNNPR	*Bench Maji*	6°47'N 35°12'E
2	C2	**Kulbisa**	SNNPR	*Bench Maji*	6°4'N 35°18'E
30	E5	**K'ulubi**	Somali	*Siti*	9°29'N 41°42'E
6	A1	**Kundu**	Oromia	*Bale*	6°23'N 39°35'E
31	G5	**K'undudo Terara**	Oromia	*East Hararge*	9°26'N 42°21'E
2	B2	**Kungul**	SNNPR	*Bench Maji*	5°54'N 34°60'E
29	I4	**Kurbe Terara**	Afar	*Gabi (Zone 3)*	9°53'N 40°26'E
2	C2	**Kurma**	SNNPR	*Bench Maji*	6°1'N 35°27'E
37	H2	**Kurub Terara**	Afar	*Awsi (Zone 1)*	11°53'N 41°13'E
15	H4	**Kus Dulamo YeTerara Senselet**	Somali	*Shabelle*	6°50'N 42°32'E
27	H5	**Kusaye Terara**	Oromia	*West Shewa*	9°24'N 37°48'E
2	B2	**Kutul**	SNNPR	*Bench Maji*	5°54'N 35°2'E
2	B2	**Kutul Birino**	SNNPR	*Bench Maji*	5°54'N 35°1'E
21	F2	**Kuyu Terara**	Amhara	*North Shewa (R3)*	8°53'N 39°39'E
35	I3	**K'want'e Terara**	Amhara	*North Wollo*	11°39'N 39°18'E

L

18	C1	**Laca**	Oromia	*East Wellega*	9°17'N 36°36'E
26	D5	**Lafto Terara**	Oromia	*Horo Guduru*	9°34'N 36°51'E
13	F3	**Lajo**	Oromia	*West Arsi*	7°12'N 39°38'E
31	F5	**Lalu Terara**	Dire Dawa	*Dire Dawa*	9°34'N 42°5'E
30	B2	**Langudi**	Afar	*Gabi (Zone 3)*	10°37'N 41°2'E
34	B5	**Lechema**	Amhara	*West Gojam*	10°57'N 37°17'E
36	B4	**Legda**	Amhara	*South Wollo*	11°23'N 39°54'E
25	H4	**Lein**	Benishangul Gumuz	*Kemashi*	9°47'N 35°27'E
14	D4	**Lele**	Oromia	*Bale*	6°42'N 41°27'E
26	C3	**Lench'a Terara**	Oromia	*East Wellega*	9°57'N 36°34'E
34	B5	**Lich'ma**	Amhara	*West Gojam*	10°57'N 37°17'E
34	B4	**Lieu**	Amhara	*West Gojam*	11°9'N 37°19'E
40	B3	**Liga**	Amhara	*Wag Himra*	13°16'N 38°36'E
34	B4	**Lijambera**	Amhara	*West Gojam*	11°7'N 37°24'E
41	H4	**Limmo**	Afar	*Fenti (Zone 4)*	12°50'N 40°12'E
40	D3	**Lisha**	Tigray	*Southern*	13°13'N 39°6'E
21	I3	**Lochu**	Oromia	*West Hararge*	8°41'N 40°19'E
37	F2	**Loma Terara**	Afar	*Awsi (Zone 1)*	11°57'N 40°55'E
21	I3	**Lucio**	Oromia	*West Hararge*	8°44'N 40°22'E
33	G1	**Luza**	Amhara	*North Gonder*	12°11'N 36°20'E

M

39	G3	**Macatat**	Amhara	*North Gonder*	13°10'N 37°31'E
35	I2	**Machel Geras**	Amhara	*North Wollo*	12°6'N 39°19'E
23	I3	**Macoda**	Oromia	*East Hararge*	8°42'N 42°44'E
18	A3	**Madugga**	Oromia	*Ilubabor*	8°36'N 36°5'E
40	E1	**Magab**	Tigray	*Eastern*	13°52'N 39°18'E
41	H4	**Magalle**	Afar	*Kilbati (Zone 2)*	12°53'N 40°3'E
36	C4	**Maghenta**	Afar	*Awsi (Zone 1)*	11°15'N 40°10'E
41	H3	**Mahalta**	Afar	*Kilbati (Zone 2)*	13°1'N 40°13'E
11	F2	**Mai Gudo**	Oromia	*Jimma*	7°29'N 37°12'E
40	E1	**Maiu**	Tigray	*Central*	13°43'N 39°17'E
2	D1	**Maji**	SNNPR	*Bench Maji*	6°9'N 35°36'E
18	C4	**Mamme**	Oromia	*Ilubabor*	8°8'N 36°28'E
32	B4	**Manchelung**	Benishangul Gumuz	*Metekel*	11°25'N 35°5'E
37	G2	**Manda YeTerara Senselet**	Afar	*Awsi (Zone 1)*	11°52'N 41°1'E
19	G5	**Manetti**	Oromia	*Jimma*	8°6'N 37°32'E
29	G3	**Manya YeTerara Senselet**	Afar	*Hari (Zone 5)*	10°4'N 39°56'E
33	I3	**Mara**	Amhara	*West Gojam*	11°29'N 36°49'E
18	A1	**Marecce**	Oromia	*West Wellega*	9°11'N 36°2'E
31	I2	**Marmar Gedle**	Somali	*Siti*	10°30'N 42°45'E
31	G1	**Maror**	Somali	*Siti*	10°53'N 42°17'E
40	D5	**Mascalo**	Amhara	*Wag Himra*	12°32'N 39°1'E
32	E5	**Masinier**	Benishangul Gumuz	*Metekel*	11°5'N 35°48'E
12	B4	**Maskulli**	SNNPR	*Sidama*	6°37'N 38°33'E
20	E2	**Masno**	Amhara	*North Shewa (R3)*	9°8'N 39°24'E
40	E1	**Masobu**	Tigray	*Eastern*	13°49'N 39°28'E
17	F3	**Mawo Terara**	Oromia	*Kelem Wellega*	8°34'N 34°48'E
40	E1	**Mayou**	Tigray	*Eastern*	13°51'N 39°28'E
35	I2	**Mechena**	Amhara	*North Wollo*	11°58'N 39°5'E
34	E5	**Medeg**	Amhara	*East Gojam*	10°56'N 38°16'E
37	G3	**Megenta YeTerara Senselet**	Afar	*Awsi (Zone 1)*	11°28'N 41°11'E
21	F1	**Megezez**	Amhara	*North Shewa (R3)*	9°17'N 39°32'E
39	F3	**Megmna**	Amhara	*North Gonder*	13°4'N 37°10'E
34	B4	**Mekari**	Amhara	*West Gojam*	11°25'N 37°22'E
27	H2	**Mekera Kir Terara**	Amhara	*East Gojam*	10°38'N 37°46'E
31	I1	**Mekho**	Somali	*Siti*	10°48'N 42°40'E
18	A3	**Meko**	Oromia	*Ilubabor*	8°38'N 36°5'E
39	G1	**Melau**	Tigray	*Western*	13°39'N 37°25'E
34	E2	**Melza**	Amhara	*South Gonder*	12°1'N 38°12'E
29	G4	**Membret**	Amhara	*North Shewa (R3)*	9°41'N 39°48'E
37	F1	**Menda Terara**	Afar	*Fenti (Zone 4)*	12°16'N 40°42'E
13	F4	**Mendebo**	Oromia	*West Arsi*	6°50'N 39°40'E
27	I2	**Mengistiwo**	Amhara	*East Gojam*	10°35'N 38°5'E
32	B4	**Mengolf**	Benishangul Gumuz	*Metekel*	11°21'N 35°5'E
38	E5	**Meniuax**	Amhara	*North Gonder*	12°29'N 36°57'E
4	D3	**Mergele Terara**	SNNPR	*Segen Peoples'*	5°42'N 37°50'E
37	H3	**Merkele Terara**	Afar	*Awsi (Zone 1)*	11°31'N 41°15'E
16	E2	**Mero Terara**	Oromia	*Kelem Wellega*	9°0'N 34°45'E
41	F1	**Mesk'el**	Tigray	*Eastern*	13°59'N 39°40'E
39	J5	**Metsefet**	Amhara	*North Gonder*	12°32'N 38°8'E
39	G5	**Mica**	Amhara	*North Gonder*	12°35'N 37°29'E
39	I1	**Mihira'i Sheyt'an Terara**	Tigray	*North Western*	13°56'N 38°6'E
18	E4	**Mile**	Oromia	*Jimma*	8°16'N 36°53'E
31	I2	**Mil'milaki-Terari**	Somali	*Siti*	10°29'N 42°43'E
25	I2	**Mincio**	Benishangul Gumuz	*Metekel*	10°37'N 35°33'E
23	G4	**Mira**	Somali	*Nogob*	8°11'N 42°15'E
40	A5	**Mira**	Amhara	*Wag Himra*	12°29'N 38°28'E
33	I5	**Mirghi**	Amhara	*Awi/Agew*	10°59'N 36°51'E
31	I1	**Mirmir**	Somali	*Siti*	10°50'N 42°51'E
19	F4	**Misciu**	Oromia	*Jimma*	8°11'N 37°17'E
18	E3	**Misingo**	Oromia	*Jimma*	8°28'N 37°2'E
19	F2	**Miti**	Oromia	*West Shewa*	8°58'N 37°22'E
40	D2	**Mits'ah Bered Terara**	Tigray	*Southern*	13°22'N 39°14'E
34	B5	**Mizan**	Amhara	*West Gojam*	11°1'N 37°28'E
34	E3	**Mize**	Amhara	*South Gonder*	11°40'N 38°8'E
31	J5	**Moccanis**	Somali	*Fafan*	9°34'N 42°54'E
23	I3	**Mogor**	Oromia	*East Hararge*	8°34'N 42°50'E
21	I3	**Molticha Terara**	Oromia	*West Hararge*	8°37'N 40°16'E
26	B4	**Morna**	Oromia	*East Wellega*	9°56'N 36°16'E
2	B2	**Mosab Gara**	SNNPR	*Bench Maji*	5°50'N 35°2'E
33	H3	**Mosabia**	Amhara	*Awi/Agew*	11°38'N 36°33'E
41	F2	**Mosobo Terara**	Tigray	*Southern*	13°34'N 39°31'E
18	D2	**Mote Terara**	Oromia	*East Wellega*	9°7'N 36°39'E
36	B4	**Motocolo**	Amhara	*South Wollo*	11°9'N 39°47'E
29	G2	**Mount Abuye Meda**	Amhara	*North Shewa (R3)*	10°33'N 39°48'E
2	C3	**Mount Naita**	SNNPR	*Bench Maji*	5°30'N 35°18'E
20	D4	**Moye Terara**	Oromia	*Arsi*	8°9'N 39°8'E
35	J2	**Muggia**	Amhara	*North Wollo*	12°3'N 39°24'E
11	I1	**Muggo**	SNNPR	*Selti*	7°49'N 37°58'E
2	D2	**Mui**	SNNPR	*Bench Maji*	6°4'N 35°35'E
19	J1	**Mulo**	Oromia	*West Shewa*	9°18'N 38°10'E
26	D4	**Mulu Terara**	Oromia	*East Wellega*	9°43'N 36°44'E
36	D3	**Muoniu**	Afar	*Awsi (Zone 1)*	11°30'N 40°17'E
19	J4	**Mur Terara**	SNNPR	*Gurage*	8°10'N 38°17'E
3	F4	**Murel**	SNNPR	*South Omo*	5°10'N 36°9'E
36	D4	**Murtu**	Afar	*Hari (Zone 5)*	11°14'N 40°15'E
29	J4	**Mutaka**	Afar	*Gabi (Zone 3)*	9°57'N 40°35'E
22	B2	**Mutar YeTerara Senselet**	Oromia	*West Hararge*	9°0'N 40°56'E
28	E3	**Muter**	Amhara	*North Shewa (R3)*	10°3'N 39°16'E
14	D2	**Mutte Mogi**	Oromia	*Bale*	7°24'N 41°33'E

N

39	F3	**Nadir**	Amhara	*North Gonder*	13°3'N 37°19'E
41	F2	**Nahaso**	Tigray	*Eastern*	13°36'N 39°36'E
40	E1	**Nahiso Terara**	Tigray	*Eastern*	13°50'N 39°17'E
2	E5	**Nakua**	SNNPR	*South Omo*	5°6'N 35°53'E
2	B2	**Nebit**	SNNPR	*Bench Maji*	5°59'N 35°6'E
36	B2	**Nedi**	Afar	*Fenti (Zone 4)*	12°6'N 39°55'E
26	C3	**Nefruit**	Oromia	*East Wellega*	10°1'N 36°28'E
16	D3	**Negu**	Oromia	*Kelem Wellega*	8°27'N 34°25'E
40	C3	**Niway YeTerara Senselet**	Tigray	*Central*	13°15'N 38°57'E
9	H4	**Nizan**	SNNPR	*Bench Maji*	6°57'N 35°27'E
40	C1	**Nobit**	Tigray	*Central*	13°42'N 38°45'E
39	H4	**Nora**	Amhara	*North Gonder*	12°43'N 37°41'E
3	F3	**Nyalibong Korebta**	SNNPR	*South Omo*	5°40'N 36°10'E

O

39	H4	**Oaroa**	Amhara	*North Gonder*	12°52'N 37°47'E
24	E2	**Obi**	Benishangul Gumuz	*Asosa*	10°37'N 34°39'E
4	E5	**Obua**	Oromia	*Borena*	4°52'N 38°5'E
16	E3	**Ola**	Oromia	*Kelem Wellega*	8°36'N 34°40'E
38	C3	**Om Zinghir**	Amhara	*North Gonder*	13°3'N 36°25'E
38	B5	**Ombibi**	Amhara	*North Gonder*	12°30'N 36°17'E
30	B5	**Onde**	Somali	*Siti*	9°35'N 41°2'E
7	J2	**Oror**	Somali	*Afder*	6°0'N 41°51'E
37	F2	**Otale Isate Gemora**	Afar	*Awsi (Zone 1)*	12°0'N 40°49'E
10	D3	**Otta**	SNNPR	*Konta*	7°6'N 36°45'E

Q

33	G1	**Quara**	Amhara	*North Gonder*	12°18'N 36°13'E
34	D5	**Quarib**	Amhara	*East Gojam*	11°4'N 37°48'E

R

19	G2	**Rafiso**	Oromia	*West Shewa*	9°0'N 37°30'E
41	H5	**Rair**	Afar	*Fenti (Zone 4)*	12°34'N 40°3'E
40	A3	**Ras Dejen**	Amhara	*North Gonder*	13°14'N 38°22'E
6	B5	**Ridderu**	Somali	*Liben*	4°49'N 39°50'E
21	J5	**Rido Terara**	Oromia	*West Hararge*	8°3'N 40°36'E
18	C3	**Rogghe**	Oromia	*East Wellega*	8°45'N 36°29'E

S

33	I2	**Saba**	Amhara	*North Gonder*	11°50'N 36°44'E
19	F4	**Sacalla**	Oromia	*Jimma*	8°20'N 37°13'E
40	D3	**Sacca**	Tigray	*Southern*	13°0'N 39°6'E
22	C5	**Sacche**	Oromia	*Bale*	8°1'N 41°12'E
35	I4	**Saco**	Amhara	*South Wollo*	11°8'N 39°17'E
21	G3	**Saddeca**	Oromia	*East Shewa*	8°43'N 39°51'E
34	D1	**Safad**	Amhara	*South Gonder*	12°11'N 37°52'E
33	H5	**Sagado**	Amhara	*Awi/Agew*	10°59'N 36°32'E
33	I4	**Saguma**	Amhara	*Awi/Agew*	11°17'N 36°46'E

37	H5	**Saha Terara**	Afar	*Awsi (Zone 1)*	11°6'N 41°19'E
18	B4	**Saie**	Oromia	*Jimma*	8°8'N 36°10'E
31	G4	**Salala**	Somali	*Siti*	9°50'N 42°13'E
13	F3	**Salmana**	Oromia	*West Arsi*	7°5'N 39°35'E
39	F2	**Sambocce**	Tigray	*Western*	13°19'N 37°11'E
34	D2	**Samo**	Amhara	*South Gonder*	12°7'N 37°57'E
20	C4	**Sap'o Terara**	Oromia	*East Shewa*	8°12'N 38°58'E
39	F5	**Sar**	Amhara	*North Gonder*	12°36'N 37°16'E
34	A5	**Satima**	Amhara	*West Gojam*	10°57'N 37°12'E
19	F1	**Sayul Terara**	Oromia	*Horo Guduru*	9°15'N 37°11'E
40	A3	**Sazza**	Amhara	*North Gonder*	13°16'N 38°26'E
11	H1	**Scerit**	SNNPR	*Hadiya*	7°46'N 37°47'E
39	G2	**Scetan**	Tigray	*Western*	13°28'N 37°33'E
10	C4	**Sciatta**	SNNPR	*Konta*	6°45'N 36°30'E
39	G2	**Scirella**	Tigray	*Western*	13°37'N 37°32'E
39	G4	**Sdni**	Amhara	*North Gonder*	12°51'N 37°27'E
36	A2	**Sebensa**	Amhara	*North Wollo*	12°7'N 39°35'E
22	E2	**Sebro Terara**	Oromia	*East Hararge*	8°48'N 41°36'E
10	B1	**Seccia**	Oromia	*Jimma*	7°45'N 36°10'E
40	B2	**Sechenchen**	Tigray	*North Western*	13°31'N 38°36'E
9	I1	**Seima**	SNNPR	*Kaffa*	7°50'N 35°44'E
9	G5	**Semab**	Gambella	*Agnuak*	6°35'N 35°8'E
33	I3	**Sena**	Amhara	*West Gojam*	11°33'N 36°49'E
11	G2	**Sengiya Terara**	SNNPR	*Hadiya*	7°20'N 37°35'E
10	D1	**Sentama**	Oromia	*Jimma*	7°43'N 36°38'E
37	J2	**Serba Terara**	Afar	*Awsi (Zone 1)*	11°50'N 41°52'E
39	I2	**Serchebin**	Tigray	*North Western*	13°28'N 37°56'E
31	G5	**Serita Terara**	Oromia	*East Hararge*	9°30'N 42°19'E
9	J5	**Shasha**	SNNPR	*Bench Maji*	6°26'N 35°53'E
21	H5	**Sheh Adere Terara**	Oromia	*Arsi*	8°6'N 40°6'E
38	C3	**Sheh Umer**	Amhara	*North Gonder*	12°59'N 36°34'E
3	I2	**Shengema YeTerara Senselet**	SNNPR	*Gamo Gofa*	5°58'N 36°45'E
11	H2	**Shenk'ola Terara**	SNNPR	*Hadiya*	7°26'N 37°45'E
26	C5	**Shenkori Terara**	Benishangul Gumuz	*Kemashi*	9°26'N 36°24'E
31	J3	**Sheyle**	Somali	*Fafan*	10°8'N 42°57'E
19	G5	**Shiro**	Oromia	*Jimma*	8°1'N 37°34'E
4	E2	**Shole YeTerara Senselet**	SNNPR	*Gedio*	5°57'N 38°13'E
4	E1	**Shomb YeTerara Senselet**	Oromia	*Borena*	6°21'N 38°11'E
10	E4	**Shurshomo**	SNNPR	*Dawro*	6°49'N 37°1'E
10	B3	**Sido**	SNNPR	*Kaffa*	7°10'N 36°20'E
36	E1	**Sifeni**	Afar	*Fenti (Zone 4)*	12°15'N 40°27'E
33	H5	**Sigadi**	Amhara	*Awi/Agew*	10°56'N 36°40'E
35	I1	**Sigit**	Amhara	*Wag Himra*	12°11'N 39°5'E
3	H2	**Siki**	SNNPR	*South Omo*	5°55'N 36°40'E
4	C4	**Sillia**	SNNPR	*Segen Peoples'*	5°24'N 37°47'E
25	G5	**Simalla**	Oromia	*West Wellega*	9°23'N 35°8'E
17	F2	**Sira**	Oromia	*Kelem Wellega*	8°57'N 34°50'E
25	H4	**Sirba**	Benishangul Gumuz	*Kemashi*	9°54'N 35°23'E
32	E5	**Sirba**	Benishangul Gumuz	*Metekel*	10°54'N 35°47'E
25	J3	**Sirbanti Terara**	Benishangul Gumuz	*Metekel*	10°9'N 35°52'E
9	I4	**Sire**	SNNPR	*Bench Maji*	6°45'N 35°36'E
21	G3	**Sirri**	Oromia	*Arsi*	8°29'N 39°47'E
35	J2	**Socona**	Amhara	*North Wollo*	12°6'N 39°26'E
18	E2	**Socso**	Oromia	*East Wellega*	8°50'N 36°58'E
18	E2	**Soddu**	Oromia	*East Wellega*	8°54'N 36°56'E
26	B3	**Soghidda**	Oromia	*East Wellega*	10°2'N 36°15'E
19	G1	**Soke Terara**	Oromia	*West Shewa*	9°19'N 37°35'E
12	E4	**Somkeru Terara**	Oromia	*West Arsi*	6°49'N 39°16'E
5	G5	**Sonkligala**	Oromia	*Borena*	4°59'N 38°41'E
16	D2	**Sonkoy**	Oromia	*Kelem Wellega*	9°4'N 34°30'E
31	J5	**Souboul Hanfele**	Somali	*Fafan*	9°35'N 43°0'E
4	D3	**Soyema YeTerara Senselet**	SNNPR	*Segen Peoples'*	5°45'N 37°50'E
40	C5	**Su Amba**	Amhara	*Wag Himra*	12°27'N 38°57'E
20	B1	**Sululta**	Oromia	*North Shewa (R4)*	9°11'N 38°39'E
37	G5	**Sundi**	Afar	*Gabi (Zone 3)*	10°56'N 41°2'E
5	G1	**Suta**	Oromia	*Guji*	6°17'N 38°35'E

T

9	F2	**Tadi**	Gambella	*Agnuak*	7°20'N 34°49'E
39	G2	**Tagle Gheorghis**	Tigray	*Western*	13°32'N 37°29'E
26	D1	**Taia**	Amhara	*Awi/Agew*	10°51'N 36°45'E
2	C2	**Taigu**	SNNPR	*Bench Maji*	6°1'N 35°23'E
27	I2	**Talo**	Amhara	*East Gojam*	10°36'N 37°54'E
11	G3	**Tambaro**	SNNPR	*Kembata Tembaro*	7°12'N 37°32'E
9	G3	**Tangader**	SNNPR	*Bench Maji*	7°6'N 35°11'E
18	B1	**Tankara**	Benishangul Gumuz	*Kemashi*	9°9'N 36°11'E
35	J1	**Taragana**	Amhara	*North Wollo*	12°11'N 39°27'E
39	G4	**Tarat**	Amhara	*North Gonder*	12°41'N 37°24'E
29	F4	**T'armaber**	Amhara	*North Shewa (R3)*	9°53'N 39°44'E
2	C2	**Taye**	SNNPR	*Bench Maji*	6°5'N 35°20'E
21	F1	**Tedda Mariam**	Amhara	*North Shewa (R3)*	9°13'N 39°31'E
40	E3	**Tegora YeTerara Senselet**	Tigray	*Southern*	13°3'N 39°28'E
21	I3	**Tek'edeyo Terara**	Oromia	*West Hararge*	8°46'N 40°25'E
39	J1	**Tekeze Terara**	Tigray	*North Western*	13°49'N 38°11'E
4	B2	**Tempe Terara**	SNNPR	*Gamo Gofa*	6°1'N 37°20'E
24	E1	**Tenze**	Benishangul Gumuz	*Asosa*	10°39'N 34°38'E
36	A1	**Thec**	Amhara	*North Wollo*	12°18'N 39°39'E
2	D2	**Tiacki**	SNNPR	*Bench Maji*	6°6'N 35°34'E
39	H3	**Tintani**	Amhara	*North Gonder*	13°1'N 37°45'E
2	B2	**Tirmakara**	SNNPR	*Bench Maji*	6°4'N 35°9'E
5	I2	**Toba**	Oromia	*Guji*	6°1'N 39°6'E
39	H3	**Tocri**	Amhara	*North Gonder*	13°9'N 37°46'E
17	F2	**Tonghi**	Oromia	*Kelem Wellega*	8°54'N 34°50'E
39	F3	**Torat**	Amhara	*North Gonder*	13°10'N 37°13'E
22	E2	**Tortora YeTerara Senselet**	Oromia	*East Hararge*	8°58'N 41°38'E
24	C5	**T'osho Terara**	Benishangul Gumuz	*Asosa*	9°31'N 34°9'E
41	F1	**Ts'arya Terara**	Tigray	*Eastern*	13°53'N 39°38'E
40	D4	**Tsellari**	Amhara	*Wag Himra*	12°50'N 39°0'E
41	F5	**Tsoiag**	Tigray	*Southern*	12°32'N 39°35'E
38	E1	**Tucul Duma**	Tigray	*Western*	13°52'N 36°58'E
2	B3	**Tulga**	SNNPR	*Bench Maji*	5°41'N 35°8'E
19	I2	**Tuli**	Oromia	*West Shewa*	9°4'N 37°56'E
19	F1	**Tulu Amara Terara**	Oromia	*Horo Guduru*	9°11'N 37°17'E
13	G4	**Tulu Dimtu**	Oromia	*Bale*	6°50'N 39°49'E
25	F5	**Tulu Arba**	Oromia	*West Wellega*	9°23'N 34°50'E
16	D4	**Tulu Aro**	Gambella	*Agnuak*	8°25'N 34°29'E
28	A3	**Tulu Bedada**	Oromia	*North Shewa (R4)*	9°59'N 38°20'E
17	G3	**Tulu Benni**	Oromia	*Ilubabor*	8°36'N 35°11'E
16	E3	**Tulu Cachir**	Oromia	*Kelem Wellega*	8°29'N 34°33'E
13	H1	**Tulu Daltu**	Oromia	*Arsi*	7°43'N 40°12'E
17	F2	**Tulu Deibia**	Oromia	*Kelem Wellega*	9°4'N 34°57'E
25	I5	**Tulu Eriwani**	Oromia	*West Wellega*	9°27'N 35°38'E
16	D3	**Tulu Gallaciu**	Oromia	*Kelem Wellega*	8°39'N 34°30'E
34	C1	**Tulu Ghereccia**	Amhara	*North Gonder*	12°16'N 37°38'E
17	I1	**Tulu Ghinghi**	Oromia	*West Wellega*	9°21'N 35°40'E
26	E5	**Tulu Gudene Terara**	Oromia	*Horo Guduru*	9°31'N 37°5'E
16	D2	**Tulu Inghi**	Benishangul Gumuz	*Asosa*	9°0'N 34°26'E
21	F5	**Tulu Sciato**	Oromia	*Arsi*	7°53'N 39°43'E
16	E3	**Tulu Sciau**	Oromia	*Kelem Wellega*	8°33'N 34°32'E
24	E4	**Tulu Sciava**	Benishangul Gumuz	*Asosa*	9°41'N 34°38'E
24	E4	**Tulu Sida**	Benishangul Gumuz	*Asosa*	9°50'N 34°34'E
21	F3	**Tulu Uaba**	Oromia	*Arsi*	8°28'N 39°34'E
25	G4	**Tulu Uandi**	Oromia	*West Wellega*	9°41'N 35°11'E
17	H5	**Tulu Uarc**	Oromia	*Ilubabor*	8°2'N 35°29'E
17	F2	**Tulu Welel**	Oromia	*Kelem Wellega*	8°53'N 34°50'E
11	H3	**Turduna Terara**	SNNPR	*Kembata Tembaro*	7°16'N 37°40'E
2	C3	**Turmu**	SNNPR	*Bench Maji*	5°37'N 35°22'E
39	J1	**Tzembelta**	Tigray	*North Western*	13°51'N 38°15'E
40	D1	**Tzili**	Tigray	*Central*	13°43'N 39°14'E

U

18	B4	**Uace**	Oromia	*Jimma*	8°12'N 36°17'E
34	C1	**Uadagi**	Amhara	*South Gonder*	12°15'N 37°46'E
26	B4	**Uaggi**	Oromia	*East Wellega*	9°42'N 36°18'E

38	D4	**Uainia**	Amhara	*North Gonder*	12°36'N 36°47'E
34	E5	**Uajzazir**	Amhara	*East Gojam*	11°1'N 38°12'E
18	D2	**Uakorma**	Oromia	*East Wellega*	9°5'N 36°48'E
40	D3	**Ual Casciua**	Amhara	*Wag Himra*	12°59'N 38°59'E
31	H5	**Ualinsi**	Oromia	*East Hararge*	9°32'N 42°25'E
19	I3	**Uarabo**	Oromia	*South West Shewa*	8°31'N 38°5'E
27	H1	**Uari**	Amhara	*East Gojam*	10°46'N 37°43'E
10	E3	**Uarrai**	SNNPR	*Dawro*	7°3'N 37°7'E
39	I3	**Uasange**	Amhara	*North Gonder*	13°10'N 37°56'E
21	G1	**Uasil**	Amhara	*North Shewa (R3)*	9°17'N 39°47'E
41	G3	**Uassi**	Afar	*Kilbati (Zone 2)*	13°12'N 39°52'E
12	C1	**Uayu**	Oromia	*Arsi*	7°48'N 38°45'E
24	E1	**Ube**	Benishangul Gumuz	*Asosa*	10°38'N 34°40'E
20	C2	**Udama**	Oromia	*North Shewa (R4)*	9°8'N 38°53'E
4	E2	**Udicussa**	Oromia	*Borena*	5°57'N 38°9'E
19	F4	**Ueca**	Oromia	*Jimma*	8°20'N 37°17'E
34	D1	**Uenfit**	Amhara	*South Gonder*	12°13'N 37°55'E
26	B3	**Ueni**	Benishangul Gumuz	*Kemashi*	10°4'N 36°11'E
18	E2	**Uera**	Oromia	*East Wellega*	8°56'N 36°56'E
37	I1	**Umbaba**	Afar	*Awsi (Zone 1)*	12°21'N 41°42'E
11	H4	**Umbo**	SNNPR	*Wolayita*	6°46'N 37°51'E
32	D2	**Umm Zenegir**	Amhara	*North Gonder*	11°58'N 35°35'E
30	B4	**Unda Rugdaia**	Somali	*Siti*	9°39'N 41°1'E
22	A2	**Unde**	Oromia	*West Hararge*	9°1'N 40°48'E
37	H1	**Undurur YeTerara Senselet**	Afar	*Awsi (Zone 1)*	12°10'N 41°24'E
8	D3	**Ungwala**	Gambella	*Agnuak*	6°57'N 34°26'E
39	I3	**Uochen**	Amhara	*North Gonder*	13°6'N 37°56'E
36	C3	**Uodedsca**	Afar	*Awsi (Zone 1)*	11°35'N 40°3'E
5	H4	**Uodo**	Oromia	*Guji*	5°20'N 38°56'E
41	H3	**Uogare**	Afar	*Kilbati (Zone 2)*	13°13'N 40°2'E
19	G2	**Uogedi**	Oromia	*West Shewa*	9°0'N 37°28'E
39	H3	**Uoghed**	Amhara	*North Gonder*	13°10'N 37°48'E
11	G5	**Uoide**	SNNPR	*Gamo Gofa*	6°35'N 37°25'E
11	G4	**Uolamo**	SNNPR	*Wolayita*	6°50'N 37°37'E
40	E5	**Uomberat**	Tigray	*Southern*	12°27'N 39°28'E
36	C2	**Uoneds**	Afar	*Fenti (Zone 4)*	12°3'N 40°6'E
11	G5	**Uoscio**	SNNPR	*Gamo Gofa*	6°32'N 37°38'E
29	F4	**Uoti**	Amhara	*North Shewa (R3)*	9°46'N 39°43'E
13	F3	**Urgoma**	Oromia	*West Arsi*	6°59'N 39°37'E
41	J4	**Urikomam Terara**	Afar	*Fenti (Zone 4)*	12°53'N 40°34'E
27	I1	**Urrata**	Amhara	*East Gojam*	10°40'N 37°58'E
27	J4	**Uru**	Oromia	*North Shewa (R4)*	9°45'N 38°15'E
36	D3	**Uruyegifi Terara**	Afar	*Awsi (Zone 1)*	11°47'N 40°22'E
12	C4	**Ussota**	Oromia	*West Arsi*	6°50'N 38°55'E
19	F5	**Utubo**	Oromia	*Jimma*	8°3'N 37°10'E
35	J4	**Uwa Terara**	Amhara	*South Wollo*	11°10'N 39°25'E

V

10	D2	**Vennio**	Oromia	*Jimma*	7°25'N 36°45'E
15	H4	**Viyeleh**	Somali	*Shabelle*	6°39'N 42°31'E

W

41	F3	**Waga'ade Terara**	Tigray	*Southern*	13°4'N 39°41'E
40	E3	**Walk'e Terara**	Tigray	*Southern*	13°10'N 39°20'E
27	F2	**Wamit**	Amhara	*East Gojam*	10°18'N 37°11'E
20	A2	**Wata Dalecha Terara**	Oromia	*West Shewa*	8°57'N 38°28'E
18	C1	**Wata Terara**	Oromia	*East Wellega*	9°20'N 36°27'E
20	B2	**Wech'ech'a YeTerara Senselet**	Oromia	*West Shewa*	8°59'N 38°35'E
31	F5	**Wedaja Terara**	Dire Dawa	*Dire Dawa*	9°35'N 42°3'E
19	H2	**Wedesa**	Oromia	*West Shewa*	9°2'N 37°47'E
35	J2	**Wek'o Terara**	Amhara	*North Wollo*	11°56'N 39°25'E
30	E4	**Welale Terara**	Somali	*Siti*	9°52'N 41°43'E
21	G4	**Welargi Terara**	Oromia	*Arsi*	8°19'N 39°58'E
21	H2	**Weldoyi Terara**	Oromia	*West Hararge*	8°49'N 40°8'E
41	F2	**Wele'a Terara**	Tigray	*Southern*	13°21'N 39°44'E
20	C3	**Werabecha Terara**	Oromia	*East Shewa*	8°41'N 38°56'E
31	G5	**Were Mucha Terara**	Oromia	*East Hararge*	9°31'N 42°11'E
36	B4	**Wereba Abo YeTerara Senselet**	Amhara	*South Wollo*	11°16'N 39°48'E
19	I2	**Werk'e**	Oromia	*West Shewa*	9°5'N 38°8'E
28	E4	**Weyara Amba**	Amhara	*North Shewa (R3)*	9°57'N 39°16'E
40	A3	**Weynobar (Ancua)**	Amhara	*North Gonder*	13°16'N 38°25'E
41	G2	**Womberta**	Tigray	*Eastern*	13°36'N 39°50'E

Y

4	E4	**Yabal**	Oromia	*Borena*	5°17'N 38°6'E
26	B4	**Yaso YeTerara Senselet**	Oromia	*East Wellega*	9°52'N 36°13'E
34	E5	**Yecandach Mountains**	Amhara	*East Gojam*	11°5'N 38°15'E
36	B5	**Yegof Terara**	Amhara	*South Wollo*	11°2'N 39°42'E
9	G2	**Yina Terara**	SNNPR	*Sheka*	7°36'N 35°17'E

Z

39	F3	**Zagol**	Amhara	*North Gonder*	13°17'N 37°17'E
40	A1	**Zana**	Tigray	*North Western*	13°53'N 38°25'E
39	G1	**Zana Terara**	Tigray	*Western*	13°42'N 37°23'E
28	E3	**Zaro Terara**	Amhara	*North Shewa (R3)*	9°58'N 39°20'E
27	I2	**Zaua**	Amhara	*East Gojam*	10°34'N 37°53'E
16	D1	**Zawa Terara**	Oromia	*West Wellega*	9°19'N 34°28'E
39	F1	**Zeba Uonti**	Tigray	*Western*	13°54'N 37°7'E
40	B1	**Zebo**	Tigray	*Central*	13°59'N 38°38'E
39	F2	**Zelal**	Tigray	*Western*	13°36'N 37°18'E
39	G2	**Zerbabit**	Tigray	*Western*	13°31'N 37°30'E
40	E1	**Zeri`a Awro YeTerara Ch'af**	Tigray	*Eastern*	13°54'N 39°22'E
40	C1	**Ziban Amba**	Tigray	*Central*	13°59'N 38°51'E
20	C3	**Zik'wala Terara**	Oromia	*East Shewa*	8°33'N 38°52'E
36	B1	**Zobil**	Amhara	*North Wollo*	12°15'N 39°45'E
39	J5	**Zoz**	Amhara	*North Gonder*	12°24'N 38°10'E
39	J2	**Zufan**	Amhara	*North Gonder*	13°20'N 38°13'E

Towns/Settlements

A

10	E3	**Aba**	SNNPR	*Dawro*	7°10'N 36°54'E
18	B4	**Aba Bor**	Oromia	*Ilubabor*	8°25'N 36°7'E
21	G3	**Abadir**	Oromia	*East Shewa*	8°45'N 39°54'E
18	B4	**Abdela**	Oromia	*Ilubabor*	8°22'N 36°14'E
38	C1	**Abderafi**	Tigray	*Western*	13°42'N 36°28'E
12	A4	**Abela Lida**	SNNPR	*Sidama*	6°56'N 38°28'E
19	G4	**Abelti**	Oromia	*Jimma*	8°10'N 37°33'E
40	D3	**Aberqele**	Tigray	*Southern*	13°2'N 39°59'E
40	D2	**Abiy Adi**	Tigray	*Central*	13°37'N 38°59'E
8	E1	**Abobo**	Gambella	*Agnuak*	7°54'N 34°32'E
21	G3	**Abomsa**	Oromia	*Arsi*	8°29'N 39°49'E
20	B5	**Abosa**	Oromia	*East Shewa*	8°1'N 38°43'E
27	F4	**Achane**	Oromia	*Horo Guduru*	9°39'N 37°19'E
27	I1	**Achewa Bado**	Amhara	*East Gojam*	10°49'N 38°1'E
26	E1	**Achigi**	Amhara	*West Gojam*	10°43'N 36°57'E
12	E3	**Adaba**	Oromia	*West Arsi*	7°0'N 39°23'E
20	E3	**Adama (Nazret)**	Oromia	*East Shewa*	8°32'N 39°16'E
18	C2	**Adami**	Oromia	*East Wellega*	9°5'N 36°28'E
12	B1	**Adami Tulu**	Oromia	*East Shewa*	7°51'N 38°42'E
20	C2	**ADDIS ABABA**	Addis Ababa	*Region 14*	9°1'N 38°45'E
13	G1	**Adele**	Oromia	*Arsi*	7°46'N 39°54'E
34	B4	**Adet**	Amhara	*West Gojam*	11°15'N 37°29'E
39	I2	**Adi Arkay**	Amhara	*North Gonder*	13°27'N 38°3'E
31	F2	**Adi Gala**	Somali	*Siti*	10°27'N 41°56'E
41	F3	**Adi Gudom**	Tigray	*Southern*	13°15'N 39°31'E
39	F1	**Adi Remets**	Tigray	*Western*	13°44'N 37°19'E
41	F4	**Adi Shoh**	Tigray	*Southern*	12°51'N 39°33'E
39	G3	**Adi Teweld Medhin**	Amhara	*North Gonder*	13°17'N 37°23'E
40	D3	**Adi Zeyla**	Tigray	*Southern*	13°6'N 39°1'E
31	G2	**Adigala**	Somali	*Siti*	10°25'N 42°14'E
20	A2	**Adis Alem**	Oromia	*West Shewa*	9°2'N 38°24'E
21	F3	**Adis Hiywet**	Oromia	*Arsi*	8°31'N 39°35'E
33	I5	**Adis Kidane**	Amhara	*Awi/Agew*	11°5'N 36°53'E
34	C2	**Adis Zemen**	Amhara	*South Gonder*	12°7'N 37°46'E
20	C3	**Adulala**	Oromia	*East Shewa*	8°32'N 38°54'E
30	B5	**Afdem**	Somali	*Siti*	9°28'N 40°59'E
13	G3	**Agarfa**	Oromia	*Bale*	7°16'N 39°49'E
10	C1	**Agaro**	Oromia	*Jimma*	7°51'N 36°35'E
26	E3	**Agemsa**	Oromia	*Horo Guduru*	10°8'N 36°56'E

19	I4	**Agena**	SNNPR	*Gurage*	8°08'N 38°00'E
4	E3	**Agere Mariyam**	Oromia	*Borena*	5°38'N 38°14'E
18	C5	**Ageyo**	Oromia	*Ilubabor*	8°2'N 36°27'E
34	C4	**Agta**	Amhara	*West Gojam*	11°20'N 37°34'E
41	F1	**Agula**	Tigray	*Eastern*	13°41'N 39°35'E
33	J5	**Agut**	Amhara	*West Gojam*	10°50'N 37°9'E
12	A3	**Aje**	Oromia	*West Arsi*	7°17'N 38°21'E
35	G5	**Ajibar**	Amhara	*South Wollo*	10°51'N 38°35'E
20	C2	**Akaki Beseka**	Addis Ababa	*Region 14*	8°52'N 38°47'E
41	F5	**Alamata**	Tigray	*Southern*	12°24'N 39°33'E
20	D1	**Aleltu**	Oromia	*North Shewa (R4)*	9°11'N 39°9'E
11	J2	**Alem Gebeya**	SNNPR	*Selti*	7°34'N 38°11'E
20	B2	**Alem Gena**	Oromia	*South West Shewa*	8°55'N 38°39'E
28	D3	**Alem Ketema**	Amhara	*North Shewa (R3)*	10°3'N 39°0'E
31	F5	**Alem Maya**	Oromia	*East Hararge*	9°23'N 42°1'E
17	G2	**Alem Teferi**	Oromia	*Kelem Wellega*	8°54'N 35°13'E
20	C4	**Alem Tena**	Oromia	*East Shewa*	8°17'N 38°56'E
12	A5	**Aleta Wendo**	SNNPR	*Sidama*	6°36'N 38°25'E
18	E3	**Alga**	Oromia	*Jimma*	8°35'N 37°6'E
17	I3	**Alge**	Oromia	*Ilubabor*	8°35'N 35°44'E
13	G3	**Ali**	Oromia	*Bale*	7°14'N 39°55'E
28	B4	**Ali Doro**	Oromia	*North Shewa (R4)*	9°46'N 38°31'E
26	E4	**Alibo**	Oromia	*Horo Guduru*	9°53'N 37°4'E
29	G5	**Aliyu Amba**	Amhara	*North Shewa (R3)*	9°34'N 39°48'E
5	G3	**Alona**	Oromia	*Guji*	5°45'N 38°40'E
27	G2	**Amanuel**	Amhara	*East Gojam*	10°26'N 37°34'E
39	G4	**Amba Giyorgis**	Amhara	*North Gonder*	12°46'N 37°37'E
35	H3	**Amba Lamba**	Amhara	*North Wollo*	11°41'N 39°2'E
41	F4	**Ambalage**	Tigray	*Southern*	12°56'N 39°31'E
19	H2	**Ambo**	Oromia	*West Shewa*	8°58'N 37°51'E
18	E5	**Ambuye**	Oromia	*Jimma*	7°58'N 36°54'E
40	B5	**Amde Werk**	Amhara	*Wag Himra*	12°26'N 38°43'E
34	D2	**Amed Ber**	Amhara	*South Gonder*	11°54'N 37°53'E
10	D3	**Ameya**	SNNPR	*Konta*	7°6'N 36°40'E
33	I2	**Amistiya**	Amhara	*North Gonder*	12°1'N 36°55'E
34	C3	**Anbesame**	Amhara	*South Gonder*	11°42'N 37°37'E
36	B5	**Ancharo**	Amhara	*South Wollo*	11°2'N 39°46'E
28	E4	**Anchekorer**	Amhara	*North Shewa (R3)*	9°48'N 39°17'E
10	B3	**Anderacha**	SNNPR	*Kaffa*	7°12'N 36°16'E
21	G4	**Angada**	Oromia	*Arsi*	8°23'N 39°49'E
11	H3	**Angecha**	SNNPR	*Kembata Tembaro*	7°12'N 37°46'E
27	G1	**Anjeni**	Amhara	*West Gojam*	10°40'N 37°32'E
29	F5	**Ankober**	Amhara	*North Shewa (R3)*	9°35'N 39°44'E
18	E2	**Ano**	Oromia	*East Wellega*	9°5'N 36°57'E
12	A4	**Aposto**	SNNPR	*Sidama*	6°44'N 38°22'E
31	I4	**Arabi**	Somali	*Siti*	9°55'N 42°42'E
34	C3	**Arb Gebeya**	Amhara	*South Gonder*	11°38'N 37°44'E
4	B2	**Arba Minch**	SNNPR	*Gamo Gofa*	6°1'N 37°33'E
22	B2	**Arba Reketi**	Oromia	*West Hararge*	9°3'N 40°54'E
12	B4	**Arbe Gona**	SNNPR	*Sidama*	6°41'N 38°43'E
21	F4	**Arboye**	Oromia	*Arsi*	8°22'N 39°40'E
11	H3	**Areka**	SNNPR	*Wolayita*	7°4'N 37°42'E
20	E2	**Arerti**	Amhara	*North Shewa (R3)*	8°55'N 39°25'E
4	B3	**Arguba**	SNNPR	*Segen Peoples'*	5°44'N 37°23'E
18	B2	**Arjo**	Oromia	*East Wellega*	9°2'N 36°14'E
18	C3	**Arjo**	Oromia	*East Wellega*	8°45'N 36°30'E
19	I5	**Arkit**	SNNPR	*Gurage*	8°0'N 38°6'E
29	G4	**Armanaya**	Amhara	*North Shewa (R3)*	9°52'N 39°49'E
12	B2	**Arsi Negele**	Oromia	*West Arsi*	7°21'N 38°40'E
21	F1	**Asagirt**	Amhara	*North Shewa (R3)*	9°24'N 39°35'E
37	H3	**Asayita**	Afar	*Awsi (Zone 1)*	11°34'N 41°26'E
22	B2	**Asbe Teferi**	Oromia	*West Hararge*	9°4'N 40°51'E
30	C3	**Asbuli**	Somali	*Siti*	9°58'N 41°9'E
11	F1	**Asendabo**	Oromia	*Jimma*	7°45'N 37°13'E
16	E3	**Ashi (Sedi)**	Oromia	*Kelem Wellega*	8°39'N 34°42'E
40	D5	**Asketema**	Amhara	*Wag Himra*	12°24'N 39°0'E
24	E3	**ASOSA**	Benishangul Gumuz	*Asosa*	10°3'N 34°32'E
12	D3	**Assasa**	Oromia	*West Arsi*	7°6'N 39°11'E
20	D5	**Assela**	Oromia	*Arsi*	7°57'N 39°7'E
29	G2	**Ataye**	Amhara	*North Shewa (R3)*	10°21'N 39°56'E
18	E4	**Atnago**	Oromia	*Jimma*	8°18'N 36°56'E
41	F1	**Atsbi Inda Silase**	Tigray	*Eastern*	13°51'N 39°43'E
21	H2	**Awash**	Afar	*Gabi (Zone 3)*	8°59'N 40°9'E
26	D4	**Ayana**	Oromia	*East Wellega*	9°52'N 36°37'E
30	D4	**Aydora**	Somali	*Siti*	9°51'N 41°22'E
31	H1	**Ayisha**	Somali	*Siti*	10°45'N 42°34'E
38	E5	**Aykel**	Amhara	*North Gonder*	12°30'N 37°2'E
35	H1	**Ayna**	Amhara	*North Wollo*	12°9'N 38°51'E
26	D1	**Azena**	Amhara	*Awi/Agew*	10°45'N 36°48'E
39	G5	**Azezo**	Amhara	*North Gonder*	12°32'N 37°25'E

B

23	G1	**Babile**	Oromia	*East Hararge*	9°12'N 42°21'E
18	E4	**Babo**	Oromia	*Jimma*	8°10'N 36°55'E
9	J4	**Bachuma**	SNNPR	*Bench Maji*	6°49'N 35°54'E
34	B3	**BAHIR DAR**	Amhara	*West Gojam*	11°35'N 37°22'E
18	E2	**Bako**	Oromia	*West Shewa*	9°7'N 37°3'E
3	H2	**Bako**	SNNPR	*South Omo*	5°50'N 36°37'E
20	E2	**Balchi**	Amhara	*North Shewa (R3)*	8°54'N 39°22'E
24	E4	**Bambesi**	Benishangul Gumuz	*Asosa*	9°45'N 34°43'E
18	C2	**Bandiro**	Oromia	*East Wellega*	8°55'N 36°35'E
20	A3	**Bantu**	Oromia	*South West Shewa*	8°36'N 38°22'E
36	C4	**Bati**	Amhara	*Oromia*	11°11'N 40°1'E
9	H4	**Bebeka**	SNNPR	*Bench Maji*	6°51'N 35°20'E
17	I4	**Becho**	Oromia	*Ilubabor*	8°12'N 35°41'E
10	D3	**Bedele**	Oromia	*Ilubabor*	8°27'N 36°21'E
22	E2	**Bedeno**	Oromia	*East Hararge*	9°6'N 41°37'E
22	A2	**Bedesa**	Oromia	*West Hararge*	8°54'N 40°46'E
11	I4	**Bedesa**	SNNPR	*Wolayita*	6°52'N 37°56'E
18	E4	**Bege**	Oromia	*Jimma*	8°18'N 37°6'E
16	E1	**Begi**	Oromia	*West Wellega*	9°20'N 34°32'E
4	B4	**Bekawile (Konso)**	SNNPR	*Segen Peoples'*	5°20'N 37°26'E
12	E2	**Bekoji**	Oromia	*Arsi*	7°31'N 39°15'E
13	H1	**Bele**	Oromia	*Arsi*	7°43'N 39°59'E
11	G4	**Bele**	SNNPR	*Wolayita*	6°55'N 37°31'E
5	I3	**Belesa Koricha**	Oromia	*Guji*	5°47'N 39°4'E
4	A2	**Belta**	SNNPR	*Gamo Gofa*	6°3'N 37°16'E
25	F4	**Benguwa**	Oromia	*West Wellega*	9°46'N 34°54'E
41	G1	**Berahile**	Afar	*Kilbati (Zone 2)*	13°51'N 40°0'E
10	C1	**Beshash**	Oromia	*Jimma*	7°47'N 36°28'E
35	H3	**Bete Hor**	Amhara	*North Wollo*	11°40'N 38°49'E
27	H3	**Bete Nigus**	Amhara	*East Gojam*	9°58'N 37°47'E
3	I2	**Beto**	SNNPR	*Gamo Gofa*	6°3'N 36°52'E
27	J2	**Bichena**	Amhara	*East Gojam*	10°27'N 38°12'E
19	I1	**Bicho**	Oromia	*West Shewa*	9°15'N 38°1'E
6	A3	**Bidire**	Oromia	*Bale*	5°47'N 39°32'E
18	C4	**Bido**	Oromia	*Ilubabor*	8°13'N 36°26'E
30	C5	**Bike**	Somali	*Siti*	9°31'N 41°12'E
18	E1	**Bila**	Oromia	*East Wellega*	9°17'N 37°1'E
25	I5	**Bila**	Oromia	*West Wellega*	9°22'N 35°35'E
18	E2	**Bilo**	Oromia	*East Wellega*	8°53'N 37°0'E
35	I2	**Bir Afaf**	Amhara	*North Wollo*	11°51'N 39°8'E
4	C1	**Birbir**	SNNPR	*Gamo Gofa*	6°18'N 37°46'E
17	G5	**Birbirsa**	Oromia	*Ilubabor*	8°3'N 35°9'E
5	G3	**Birbirsa Bera**	Oromia	*Guji*	5°42'N 38°35'E
27	J4	**Biriti**	Oromia	*North Shewa (R4)*	9°45'N 38°16'E
9	I3	**Bita Genet**	SNNPR	*Kaffa*	7°16'N 35°46'E
6	A3	**Bitata**	Oromia	*Guji*	5°28'N 39°28'E
19	G1	**Bitile**	Oromia	*West Shewa*	9°11'N 37°28'E
31	H3	**Biye Anod**	Somali	*Siti*	10°3'N 42°31'E
31	H2	**Biye Gurgur**	Somali	*Siti*	10°23'N 42°34'E
11	I3	**Boditi**	SNNPR	*Wolayita*	6°57'N 37°51'E
23	F2	**Boku**	Oromia	*East Hararge*	9°7'N 42°5'E
4	D5	**Bolekedo**	Oromia	*Borena*	4°57'N 37°59'E
10	E2	**Boneya**	Oromia	*Jimma*	7°32'N 37°3'E
20	B2	**Boneya**	Oromia	*South West Shewa*	8°49'N 38°38'E
17	F4	**Bonga**	Gambella	*Agnuak*	8°10'N 34°50'E
10	B3	**Bonga**	SNNPR	*Kaffa*	7°16'N 36°15'E
4	B2	**Bonke Beza**	SNNPR	*Gamo Gofa*	6°3'N 37°20'E
21	I2	**Bordele**	Oromia	*West Hararge*	9°1'N 40°23'E
12	B5	**Bore**	Oromia	*Guji*	6°21'N 38°37'E
36	A4	**Boru**	Amhara	*South Wollo*	11°13'N 39°39'E
19	F4	**Botor Bocho**	Oromia	*Jimma*	8°22'N 37°16'E
19	F4	**Botor Keta**	Oromia	*Jimma*	8°22'N 37°19'E

17	J2	**Bube**	Oromia	*West Wellega*	8°50'N 35°48'E
20	B4	**Bu-i**	SNNPR	*Gurage*	8°19'N 38°32'E
21	F5	**Bulala**	Oromia	*Arsi*	7°59'N 39°34'E
12	B1	**Bulbula**	Oromia	*East Shewa*	7°43'N 38°39'E
5	A5	**Bule**	Oromia	*Guji*	6°16'N 38°27'E
12	D5	**Bulega**	SNNPR	*Sidama*	6°35'N 39°6'E
26	A2	**Bulen**	Benishangul Gumuz	*Metekel*	10°35'N 36°4'E
3	I1	**Bulki**	SNNPR	*Gamo Gofa*	6°16'N 36°48'E
12	A3	**Bura**	Oromia	*West Arsi*	7°15'N 38°29'E
26	E1	**Bure**	Amhara	*West Gojam*	10°43'N 37°4'E
17	G4	**Bure**	Oromia	*Ilubabor*	8°17'N 35°6'E
17	H4	**Burusa**	Oromia	*Ilubabor*	8°17'N 35°27'E
20	A4	**Butajira**	SNNPR	*Gurage*	8°7'N 38°22'E

C

28	E5	**Chacha**	Amhara	*North Shewa (R3)*	9°32'N 39°27'E
33	H5	**Chagni**	Amhara	*Awi/Agew*	10°57'N 36°30'E
20	C1	**Chancho**	Oromia	*North Shewa (R4)*	9°18'N 38°45'E
38	E5	**Chandiba**	Amhara	*North Gonder*	12°23'N 37°1'E
17	G2	**Chanka**	Oromia	*Kelem Wellega*	8°51'N 35°4'E
20	D2	**Chefe Donsa**	Oromia	*East Shewa*	8°58'N 39°7'E
5	H3	**Cheketa**	Oromia	*Guji*	5°39'N 38°49'E
21	H3	**Chelekko**	Oromia	*West Hararge*	8°41'N 40°14'E
4	E2	**Chelelektu**	SNNPR	*Gedio*	6°1'N 38°9'E
30	D5	**Chelenko**	Oromia	*East Hararge*	9°23'N 41°33'E
27	H3	**Chemoga**	Amhara	*East Gojam*	10°18'N 37°45'E
4	C1	**Chencha**	SNNPR	*Gamo Gofa*	6°15'N 37°34'E
39	J3	**Chennek**	Amhara	*North Gonder*	13°15'N 38°11'E
7	J4	**Chereti**	Somali	*Afder*	5°20'N 41°53'E
10	D3	**Chida**	SNNPR	*Konta*	7°10'N 36°47'E
31	H5	**Chinaksen**	Oromia	*East Hararge*	9°31'N 42°33'E
10	B1	**Chira**	Oromia	*Jimma*	7°44'N 36°15'E
10	B3	**Chiri**	SNNPR	*Kaffa*	7°7'N 36°11'E
19	I1	**Chobi**	Oromia	*West Shewa*	9°19'N 37°56'E
18	D5	**Choche**	Oromia	*Jimma*	7°57'N 36°42'E
21	G4	**Chole**	Oromia	*Arsi*	8°8'N 39°54'E
12	A4	**Chuko**	SNNPR	*Sidama*	6°39'N 38°21'E
27	H4	**Chulute**	Oromia	*West Shewa*	9°39'N 37°44'E
34	A1	**Chwahit Mender**	Amhara	*North Gonder*	12°19'N 37°13'E

D

39	H3	**Dabat**	Amhara	*North Gonder*	12°58'N 37°45'E
34	B5	**Dabi**	Amhara	*West Gojam*	10°59'N 37°25'E
40	C5	**Dahana**	Amhara	*Wag Himra*	12°27'N 38°51'E
25	G3	**Daleti**	Benishangul Gumuz	*Asosa*	10°1'N 35°13'E
28	B4	**Daleti**	Oromia	*North Shewa (R4)*	9°36'N 38°35'E
11	J5	**Dalocha**	SNNPR	*Selti*	7°47'N 38°14'E
33	I4	**Dangla**	Amhara	*Awi/Agew*	11°15'N 36°50'E
10	A2	**Dara**	SNNPR	*Kaffa*	7°18'N 35°59'E
5	G4	**Dawa**	Oromia	*Guji*	5°19'N 38°47'E
17	F2	**Dawo Kebe**	Oromia	*Kelem Wellega*	8°58'N 34°58'E
39	I3	**Debark**	Amhara	*North Gonder*	13°9'N 37°53'E
25	G5	**Debeka**	Oromia	*West Wellega*	9°28'N 35°11'E
29	I3	**Debel**	Afar	*Hari (Zone 5)*	10°15'N 40°15'E
17	I2	**Debeso**	Oromia	*West Wellega*	8°50'N 35°39'E
17	I1	**Deboko**	Oromia	*West Wellega*	9°16'N 35°41'E
28	B5	**Debra**	Oromia	*North Shewa (R4)*	9°26'N 38°38'E
29	F4	**Debre Birhan**	Amhara	*North Shewa (R3)*	9°40'N 39°31'E
20	A3	**Debre Genet**	Oromia	*South West Shewa*	8°47'N 38°19'E
28	C4	**Debre Libanos**	Oromia	*North Shewa (R4)*	9°42'N 38°50'E
27	H2	**Debre Markos**	Amhara	*East Gojam*	10°20'N 37°43'E
21	H4	**Debre Selam**	Oromia	*Arsi*	8°15'N 40°08'E
29	G4	**Debre Sina**	Amhara	*North Shewa (R3)*	9°50'N 39°45'E
34	D2	**Debre Tabor**	Amhara	*South Gonder*	11°51'N 38°0'E
28	C4	**Debre Tsige**	Oromia	*North Shewa (R4)*	9°38'N 38°50'E
27	J1	**Debre Werk**	Amhara	*East Gojam*	10°39'N 38°9'E
35	G2	**Debre Zebit**	Amhara	*North Wollo*	11°49'N 38°36'E
25	I2	**Debre Zeyit (Wembera)**	Benishangul Gumuz	*Metekel*	10°37'N 35°39'E
20	D3	**Debre Zeyt (Bishoftu)**	Oromia	*East Shewa*	8°44'N 38°59'E
37	J2	**Deday**	Afar	*Awsi (Zone 1)*	11°54'N 41°47'E
22	D5	**Deder**	Oromia	*East Hararge*	9°18'N 41°26'E
4	D4	**Dedertu**	Oromia	*Borena*	5°13'N 38°2'E
10	D2	**Dedo**	Oromia	*Jimma*	7°30'N 36°52'E
27	G4	**Dedu**	Oromia	*Horo Guduru*	9°41'N 37°32'E
18	A3	**Dega**	Oromia	*Ilubabor*	8°34'N 36°6'E
12	C2	**Degaga**	Oromia	*West Arsi*	7°25'N 38°48'E
23	J5	**Degeh Medo**	Somali	*Jarar*	7°59'N 43°0'E
28	B4	**Degem**	Oromia	*North Shewa (R4)*	9°48'N 38°37'E
27	J3	**Dejen**	Amhara	*East Gojam*	10°10'N 38°9'E
33	J1	**Delgi**	Amhara	*North Gonder*	12°11'N 37°2'E
27	G2	**Dembecha**	Amhara	*West Gojam*	10°34'N 37°29'E
18	C5	**Dembi**	Oromia	*Ilubabor*	8°5'N 36°27'E
17	F3	**Dembidolo**	Oromia	*Kelem Wellega*	8°32'N 34°48'E
31	I4	**Denbel**	Somali	*Fafan*	9°52'N 42°45'E
28	D4	**Deneba**	Amhara	*North Shewa (R3)*	9°45'N 39°11'E
11	F1	**Deneba**	Oromia	*Jimma*	7°50'N 37°21'E
38	E2	**Densha**	Tigray	*Western*	13°33'N 36°58'E
20	E4	**Dera**	Oromia	*Arsi*	8°19'N 39°19'E
39	J3	**Deresge**	Amhara	*North Gonder*	13°4'N 38°8'E
10	B2	**Deri**	SNNPR	*Kaffa*	7°23'N 36°17'E
36	A4	**Dese**	Amhara	*South Wollo*	11°7'N 39°38'E
39	I3	**Dib Bahir**	Amhara	*North Gonder*	13°14'N 37°53'E
26	B1	**Dibate**	Benishangul Gumuz	*Metekel*	10°46'N 36°15'E
23	F5	**Didimtu**	Oromia	*East Hararge*	9°21'N 42°2'E
12	E1	**Digelu**	Oromia	*Arsi*	7°45'N 39°15'E
27	H1	**Digo Tsiyon**	Amhara	*East Gojam*	10°51'N 37°44'E
15	I3	**Dihun**	Somali	*Nogob*	7°13'N 42°41'E
19	I3	**Dilela**	Oromia	*South West Shewa*	8°38'N 38°3'E
40	A3	**Dilibza**	Amhara	*North Gonder*	13°6'N 38°26'E
12	A5	**Dilla**	SNNPR	*Gedio*	6°24'N 38°18'E
11	I4	**Dimtu**	SNNPR	*Wolayita*	6°53'N 38°1'E
29	F4	**Dinbaro**	Amhara	*North Shewa (R3)*	9°42'N 39°38'E
21	G3	**Dino**	Oromia	*Arsi*	8°40'N 39°56'E
33	J2	**Dinqil Ber**	Amhara	*North Gonder*	11°58'N 36°59'E
13	G3	**Dinshu**	Oromia	*Bale*	7°6'N 39°46'E
26	D4	**Dire**	Oromia	*East Wellega*	9°55'N 36°40'E
31	F4	**DIRE DAWA**	Dire Dawa	*Dire Dawa*	9°35'N 41°51'E
19	G4	**Dire Godu**	Oromia	*West Shewa*	8°26'N 37°25'E
12	D3	**Dodola**	Oromia	*West Arsi*	6°58'N 39°10'E
28	E2	**Dogolo**	Amhara	*South Wollo*	10°25'N 39°15'E
28	C5	**Duber**	Oromia	*North Shewa (R4)*	9°27'N 38°52'E
37	G3	**Dubti**	Afar	*Awsi (Zone 1)*	11°43'N 41°5'E
20	B4	**Dugda**	SNNPR	*Gurage*	8°10'N 38°38'E
20	C3	**Dukem**	Oromia	*East Shewa*	8°47'N 38°54'E
29	G5	**Dulecha**	Afar	*Gabi (Zone 3)*	9°32'N 39°57'E
17	H3	**Dupa**	Oromia	*Ilubabor*	8°34'N 35°31'E
11	I3	**Durame**	SNNPR	*Kembata Tembaro*	7°14'N 37°54'E
33	J4	**Durbete**	Amhara	*West Gojam*	11°21'N 36°57'E
11	G3	**Durgi**	SNNPR	*Kembata Tembaro*	7°16'N 37°28'E
10	B1	**Dusta**	Oromia	*Jimma*	7°45'N 36°11'E

E

29	G2	**Efeson**	Amhara	*North Shewa (R3)*	10°20'N 39°57'E
20	E3	**Ejere**	Oromia	*East Shewa*	8°46'N 39°15'E
28	B4	**Ejere**	Oromia	*North Shewa (R4)*	9°55'N 38°30'E
17	J3	**Elemo**	Oromia	*Ilubabor*	8°27'N 35°49'E
27	G3	**Eliyas**	Amhara	*East Gojam*	10°17'N 37°27'E
30	D5	**Erer**	Somali	*Siti*	9°32'N 41°21'E

F

4	B4	**Fasha**	SNNPR	*Segen Peoples'*	5°16'N 37°21'E
18	D2	**Fechase**	Oromia	*East Wellega*	8°58'N 36°40'E
27	I1	**Felege Birhan**	Amhara	*East Gojam*	10°44'N 38°3'E
10	C3	**Felege Selam**	SNNPR	*Kaffa*	7°6'N 36°28'E
27	G1	**Feres Bet**	Amhara	*West Gojam*	10°50'N 37°35'E
28	C3	**Fetire**	Amhara	*North Shewa (R3)*	10°0'N 38°53'E
28	C4	**Fiche**	Oromia	*North Shewa (R4)*	9°46'N 38°43'E
23	G4	**Fik**	Somali	*Nogob*	8°8'N 42°18'E
35	G2	**Filakit**	Amhara	*North Wollo*	11°46'N 38°44'E
27	J3	**Filikilik**	Oromia	*North Shewa (R4)*	10°3'N 38°15'E
7	F5	**Filtu**	Somali	*Liben*	5°6'N 40°39'E
27	F5	**Fincha-a**	Oromia	*Horo Guduru*	9°33'N 37°21'E
4	E4	**Finchawa**	Oromia	*Borena*	5°24'N 38°16'E
27	F1	**Finote Selam**	Amhara	*West Gojam*	10°40'N 37°15'E

35	I3	**Firida Mareja**	Amhara	*North Wollo*	11°33'N 39°8'E
4	E2	**Fiseha Genet**	SNNPR	*Gedio*	6°4'N 38°12'E
11	G5	**Fofa**	SNNPR	*Yem*	7°50'N 37°31'E
33	J5	**Fogota**	Amhara	*Awi/Agew*	11°3'N 37°2'E
11	I2	**Fonko**	SNNPR	*Hadiya*	7°38'N 37°57'E
23	H1	**Funyan Bira**	Oromia	*East Hararge*	9°21'N 42°23'E

G

9	J4	**Gachit**	SNNPR	*Bench Maji*	6°45'N 35°47'E
26	B2	**Galesa**	Benishangul Gumuz	*Metekel*	10°26'N 36°8'E
16	E4	**Gambela**	Gambella	*Agnuak*	8°15'N 34°35'E
16	E3	**Garigeda**	Oromia	*Kelem Wellega*	8°46'N 34°39'E
34	E3	**Gasay**	Amhara	*South Gonder*	11°47'N 38°8'E
13	H2	**Gasera**	Oromia	*Bale*	7°22'N 40°11'E
35	H3	**Gashena**	Amhara	*North Wollo*	11°41'N 38°55'E
4	B3	**Gato**	SNNPR	*Segen Peoples'*	5°33'N 37°24'E
28	A4	**Gebre Guracha**	Oromia	*North Shewa (R4)*	9°48'N 38°25'E
9	H2	**Gecha**	SNNPR	*Sheka*	7°33'N 35°24'E
18	C4	**Gechi**	Oromia	*Ilubabor*	8°20'N 36°25'E
4	E2	**Gedeb**	SNNPR	*Gedio*	5°58'N 38°15'E
39	H4	**Gedebge**	Amhara	*North Gonder*	12°52'N 37°44'E
19	G2	**Gedo**	Oromia	*West Shewa*	9°0'N 37°26'E
21	J2	**Gelemso**	Oromia	*West Hararge*	8°48'N 40°30'E
26	C4	**Gelila**	Oromia	*East Wellega*	9°50'N 36°28'E
10	D1	**Gembe**	Oromia	*Jimma*	7°50'N 36°40'E
11	I3	**Gemesha**	SNNPR	*Kembata Tembaro*	7°14'N 37°54'E
4	B4	**Gemole**	SNNPR	*Segen Peoples'*	5°17'N 37°27'E
28	D3	**Gende Bune**	Amhara	*North Shewa (R3)*	10°15'N 39°9'E
28	C1	**Genet**	Amhara	*South Wollo*	10°46'N 38°50'E
27	G1	**Genet**	Amhara	*West Gojam*	10°49'N 37°25'E
17	I2	**Genji**	Oromia	*West Wellega*	9°2'N 35°36'E
10	D3	**Genji**	SNNPR	*Konta*	7°7'N 36°45'E
18	E5	**Gent**	Oromia	*Jimma*	8°5'N 36°56'E
4	E3	**Gerba**	Oromia	*Borena*	5°46'N 38°16'E
4	A2	**Gerese**	SNNPR	*Gamo Gofa*	5°55'N 37°17'E
9	H5	**Gesha**	SNNPR	*Bench Maji*	6°33'N 35°30'E
11	G4	**Gesuba**	SNNPR	*Wolayita*	6°47'N 37°33'E
18	C2	**Getema**	Oromia	*East Wellega*	8°53'N 36°29'E
4	A4	**Gewada**	SNNPR	*Segen Peoples'*	5°25'N 37°14'E
29	J3	**Gewane**	Afar	*Gabi (Zone 3)*	10°9'N 40°39'E
11	F1	**Gibe**	Oromia	*Jimma*	7°45'N 37°11'E
39	I3	**Gich**	Amhara	*North Gonder*	13°16'N 38°6'E
16	E2	**Gidami**	Oromia	*Kelem Wellega*	8°59'N 34°36'E
4	I5	**Gidicho**	Oromia	*Borena*	6°22'N 37°57'E
4	B3	**Gidole**	SNNPR	*Segen Peoples'*	5°38'N 37°20'E
17	J1	**Gimbi**	Oromia	*West Wellega*	9°10'N 35°50'E
11	G2	**Gimbichu**	SNNPR	*Hadiya*	7°26'N 37°37'E
12	E2	**Gimbite**	Oromia	*Arsi*	7°20'N 39°22'E
10	B2	**Gimbo**	SNNPR	*Kaffa*	7°23'N 36°13'E
33	I5	**Gimja Bet**	Amhara	*Awi/Agew*	10°51'N 36°53'E
21	F1	**Gina Ager**	Amhara	*North Shewa (R3)*	9°20'N 39°34'E
19	I2	**Ginchi**	Oromia	*West Shewa*	9°1'N 38°7'E
14	A3	**Ginir**	Oromia	*Bale*	7°8'N 40°42'E
22	E2	**Girawa**	Oromia	*East Hararge*	9°8'N 41°50'E
34	A5	**Gish Abay**	Amhara	*West Gojam*	10°59'N 37°13'E
35	J3	**Gishen Maryam**	Amhara	*South Wollo*	11°31'N 39°21'E
25	F1	**Gizen**	Benishangul Gumuz	*Asosa*	10°42'N 34°46'E
13	G3	**Goba**	Oromia	*Bale*	7°1'N 39°58'E
19	F1	**Goben**	Oromia	*Horo Guduru*	9°10'N 37°18'E
13	F2	**Gobesa**	Oromia	*Arsi*	7°36'N 39°29'E
4	D3	**Goche**	SNNPR	*Segen Peoples'*	5°32'N 37°54'E
25	G2	**Godere**	Benishangul Gumuz	*Kemashi*	10°26'N 35°7'E
20	D2	**Godino**	Oromia	*East Shewa*	8°52'N 39°1'E
29	F4	**Godo Beret**	Amhara	*North Shewa (R3)*	9°48'N 39°41'E
31	I3	**Gogti**	Somali	*Fafan*	10°4'N 42°50'E
19	I1	**Goja**	Oromia	*West Shewa*	9°16'N 38°3'E
19	J1	**Gola**	Oromia	*West Shewa*	9°18'N 38°11'E
18	C1	**Golo**	Oromia	*East Wellega*	9°10'N 36°27'E
21	H4	**Gololcha**	Oromia	*Arsi*	8°11'N 40°3'E
26	D2	**Gomer**	Amhara	*West Gojam*	10°25'N 36°51'E
20	D5	**Gonde**	Oromia	*Arsi*	8°1'N 39°10'E
39	G4	**Gonder**	Amhara	*North Gonder*	12°36'N 37°28'E
34	C4	**Gonj**	Amhara	*West Gojam*	11°13'N 37°40'E
17	H4	**Gore**	Oromia	*Ilubabor*	8°9'N 35°32'E
28	C5	**Gorfo**	Oromia	*North Shewa (R4)*	9°23'N 38°50'E
34	B1	**Gorgora**	Amhara	*North Gonder*	12°14'N 37°18'E
13	J4	**Goro**	Oromia	*Bale*	6°57'N 40°37'E
27	J3	**Gosha Tsiyon**	Oromia	*North Shewa (R4)*	10°0'N 38°14'E
30	C5	**Gota**	Somali	*Siti*	9°31'N 41°19'E
10	D3	**Gowa**	SNNPR	*Konta*	7°2'N 36°39'E
32	C4	**Guba**	Benishangul Gumuz	*Metekel*	11°16'N 35°17'E
19	H4	**Gubre**	SNNPR	*Gurage*	8°10'N 37°48'E
19	H2	**Guder**	Oromia	*West Shewa*	8°58'N 37°45'E
12	B3	**Guguma**	SNNPR	*Sidama*	6°58'N 38°42'E
39	I5	**Guhala**	Amhara	*North Gonder*	12°24'N 38°4'E
17	J2	**Guji**	Oromia	*West Wellega*	9°0'N 35°49'E
17	H1	**Guliso**	Oromia	*West Wellega*	9°9'N 35°28'E
19	H5	**Gunchire**	SNNPR	*Gurage*	8°1'N 37°50'E
34	E5	**Gunde Weyn**	Amhara	*East Gojam*	10°55'N 38°5'E
28	B3	**Gundo Meskel**	Oromia	*North Shewa (R4)*	10°13'N 38°41'E
9	G4	**Gura Ferda**	SNNPR	*Bench Maji*	6°49'N 35°6'E
18	D2	**Gute**	Oromia	*East Wellega*	9°1'N 36°37'E
18	B3	**Guto**	Oromia	*Ilubabor*	8°38'N 36°10'E
34	D1	**Gwaliya**	Amhara	*South Gonder*	12°10'N 37°59'E

H

13	G1	**Habe**	Oromia	*Arsi*	7°49'N 39°47'E
14	D2	**Hada**	Oromia	*Bale*	7°21'N 41°21'E
23	I1	**Hadew**	Somali	*Fafan*	9°22'N 42°40'E
12	B5	**Hagere Selam**	SNNPR	*Sidama*	6°29'N 38°31'E
40	D2	**Hagere Selam**	Tigray	*Central*	13°39'N 39°10'E
13	F3	**Hako**	Oromia	*West Arsi*	7°9'N 39°32'E
21	F5	**Hamda Diskis**	Oromia	*Arsi*	8°3'N 39°33'E
18	A2	**Hamus Gebeya**	Oromia	*Ilubabor*	9°3'N 36°6'E
19	G2	**Hamus Gebeya**	Oromia	*West Shewa*	9°7'N 37°31'E
34	C3	**Hamusit Mender**	Amhara	*South Gonder*	11°46'N 37°33'E
3	F1	**Hana**	SNNPR	*South Omo*	6°13'N 36°7'E
23	G1	**HARAR**	Hareri	*Hareri*	9°18'N 42°7'E
31	F3	**Harawa**	Somali	*Siti*	9°55'N 41°59'E
22	C1	**Harawacha**	Oromia	*East Hararge*	9°11'N 41°20'E
36	B5	**Harbu**	Amhara	*South Wollo*	10°56'N 39°46'E
5	J3	**Hare Kelo**	Oromia	*Guji*	5°33'N 39°23'E
26	B4	**Haro**	Oromia	*East Wellega*	9°50'N 36°18'E
12	B3	**HAWASSA**	SNNPR	*Sidama*	7°2'N 38°29'E
36	B4	**Hayk**	Amhara	*South Wollo*	11°18'N 39°40'E
37	I2	**Hayu**	Afar	*Awsi (Zone 1)*	11°55'N 41°33'E
17	I1	**Hena**	Oromia	*West Wellega*	9°21'N 35°41'E
28	D2	**Hibino**	Amhara	*South Wollo*	10°27'N 39°0'E
40	E3	**Hintalo**	Tigray	*Southern*	13°14'N 39°25'E
22	C1	**Hirna**	Oromia	*West Hararge*	9°13'N 41°6'E
40	F3	**Hiwane**	Tigray	*Southern*	13°5'N 39°29'E
12	B4	**Hogiso**	SNNPR	*Sidama*	6°50'N 38°40'E
20	B2	**Holota**	Oromia	*West Shewa*	9°4'N 38°29'E
11	H2	**Hosaina**	SNNPR	*Hadiya*	7°32'N 37°51'E
27	J4	**Hose**	Oromia	*North Shewa (R4)*	9°56'N 38°10'E
22	B2	**Hunde Lafto**	Oromia	*West Hararge*	9°7'N 40°58'E
30	E4	**Hurso**	Somali	*Siti*	9°36'N 41°38'E
17	I4	**Hurumu**	Oromia	*Ilubabor*	8°20'N 35°41'E

I

34	E2	**Ibnat**	Amhara	*South Gonder*	12°7'N 38°2'E
19	F2	**Ijaji**	Oromia	*West Shewa*	9°0'N 37°19'E
15	G5	**Imi**	Somali	*Shabelle*	6°28'N 42°10'E
17	I1	**Inango**	Oromia	*West Wellega*	9°9'N 35°41'E
20	A1	**Inchini**	Oromia	*West Shewa*	9°18'N 38°22'E
39	J1	**Inda Baguna**	Tigray	*North Western*	13°57'N 38°11'E
19	I4	**Indibir**	SNNPR	*Gurage*	8°7'N 37°55'E
28	D4	**Inewari**	Amhara	*North Shewa (R3)*	9°53'N 39°8'E
27	F2	**Inewend**	Amhara	*West Gojam*	10°23'N 37°12'E
33	I5	**Injibara**	Amhara	*Awi/Agew*	10°58'N 36°55'E
19	G2	**Irensa**	Oromia	*West Shewa*	8°58'N 37°34'E
16	C4	**Itang**	Gambella	*Agnuak*	8°12'N 34°16'E
20	E4	**Iteya**	Oromia	*Arsi*	8°8'N 39°14'E

J

11	H2	**Jajura**	SNNPR	*Hadiya*	7°27'N 37°41'E
26	D3	**Jangir**	Oromia	*East Wellega*	10°4'N 36°40'E
14	A2	**Jara**	Oromia	*Bale*	7°27'N 40°48'E
34	D4	**Jara Gedo**	Amhara	*South Gonder*	11°23'N 37°52'E
26	E4	**Jardega**	Oromia	*Horo Guduru*	9°48'N 37°1'E
25	H5	**Jarso**	Oromia	*West Wellega*	9°27'N 35°19'E
4	D2	**Jelo**	SNNPR	*Segen Peoples'*	5°49'N 37°58'E
19	I5	**Jembero**	SNNPR	*Gurage*	8°1'N 38°2'E
27	I4	**Jemo Lefo**	Oromia	*West Shewa*	9°39'N 38°1'E
9	I4	**Jemu**	SNNPR	*Bench Maji*	6°38'N 35°46'E
29	G3	**Jewha**	Amhara	*Oromia*	10°6'N 39°58'E
27	G1	**Jiga**	Amhara	*West Gojam*	10°40'N 37°22'E
28	E3	**Jihur**	Amhara	*North Shewa (R3)*	10°0'N 39°14'E
23	I1	**Jijiga**	Somali	*Fafan*	9°20'N 42°47'E
16	B4	**Jikawo**	Gambella	*Nuer*	8°22'N 33°47'E
29	G2	**Jimate**	Amhara	*Oromia*	10°36'N 39°55'E
10	D1	**Jimma**	Oromia	*Jimma*	7°40'N 36°50'E
3	H3	**Jinka**	SNNPR	*South Omo*	5°47'N 36°33'E
18	A2	**Jogir**	Oromia	*West Wellega*	9°2'N 36°2'E

K

35	J5	**Kabe**	Amhara	*South Wollo*	10°49'N 39°28'E
27	H5	**Kachisi**	Oromia	*West Shewa*	9°35'N 37°50'E
3	H3	**Kaka**	SNNPR	*South Omo*	5°40'N 36°38'E
17	G2	**Kake**	Oromia	*Kelem Wellega*	8°55'N 35°3'E
29	G2	**Karakore**	Amhara	*North Shewa (R3)*	10°25'N 39°56'E
10	A2	**Kebechi**	SNNPR	*Kaffa*	7°27'N 36°6'E
20	A5	**Keber**	SNNPR	*Selti*	8°1'N 38°19'E
40	C4	**Kedamit**	Amhara	*Wag Himra*	12°52'N 38°51'E
17	J2	**Keki**	Oromia	*West Wellega*	9°4'N 35°49'E
4	D2	**Kele**	SNNPR	*Segen Peoples'*	5°50'N 37°53'E
28	D2	**Kelela**	Amhara	*South Wollo*	10°35'N 38°59'E
16	E1	**Kelem**	Oromia	*West Wellega*	9°9'N 34°31'E
13	F1	**Kelo**	Oromia	*Arsi*	7°43'N 39°33'E
26	C3	**Kelo**	Oromia	*East Wellega*	10°6'N 36°27'E
17	H2	**Kelo**	Oromia	*West Wellega*	8°55'N 35°21'E
4	A2	**Kemba**	SNNPR	*Gamo Gofa*	6°3'N 37°10'E
27	J5	**Kembolcha**	Oromia	*West Shewa*	9°30'N 38°12'E
29	G1	**Kemise**	Amhara	*Oromia*	10°42'N 39°51'E
18	B3	**Kemise**	Oromia	*Ilubabor*	8°28'N 36°10'E
10	B1	**Kemise**	Oromia	*Jimma*	7°47'N 36°21'E
34	D5	**Keraniya**	Amhara	*East Gojam*	10°58'N 37°58'E
5	F3	**Kercha**	Oromia	*Guji*	5°46'N 38°24'E
27	H4	**Kere Dobi**	Oromia	*West Shewa*	9°44'N 37°42'E
12	C2	**Kersa**	Oromia	*Arsi*	7°31'N 38°58'E
30	E5	**Kersa**	Oromia	*East Hararge*	9°27'N 41°52'E
33	J5	**Kesa**	Amhara	*Awi/Agew*	10°53'N 36°58'E
27	J5	**Ketket**	Oromia	*West Shewa*	9°22'N 38°12'E
18	C2	**Kewisa**	Oromia	*East Wellega*	8°56'N 36°28'E
27	G4	**Kewo**	Oromia	*Horo Guduru*	9°49'N 37°33'E
3	I3	**Key Afer**	SNNPR	*South Omo*	5°31'N 36°44'E
5	H2	**Kibre Mengist**	Oromia	*Guji*	5°53'N 38°59'E
33	H5	**Kidamaja**	Amhara	*Awi/Agew*	11°0'N 36°41'E
26	C1	**Kilaj**	Amhara	*Awi/Agew*	10°42'N 36°29'E
5	F3	**Kilenso**	Oromia	*Borena*	5°39'N 38°19'E
11	F4	**Kindo Halato**	SNNPR	*Wolayita*	6°45'N 37°20'E
26	E3	**Kiremu**	Oromia	*East Wellega*	9°57'N 36°52'E
24	E5	**Kober**	Oromia	*West Wellega*	9°28'N 34°33'E
36	A1	**Kobo**	Amhara	*North Wollo*	12°9'N 39°37'E
30	D5	**Kobo**	Oromia	*East Hararge*	9°22'N 41°28'E
5	F1	**Kofele**	Oromia	*Guji*	6°9'N 38°24'E
12	C3	**Kofele**	Oromia	*West Arsi*	7°4'N 38°47'E
20	D4	**Koka**	Oromia	*East Shewa*	8°26'N 39°1'E
35	I5	**Koke Ager**	Amhara	*South Wollo*	10°57'N 39°13'E
26	D4	**Kokofe**	Oromia	*East Wellega*	9°55'N 36°45'E
39	F5	**Kola Diba**	Amhara	*North Gonder*	12°25'N 37°20'E
4	A4	**Kolme**	SNNPR	*Segen Peoples'*	5°19'N 37°15'E
18	C2	**Kolobo**	Oromia	*East Wellega*	9°1'N 36°27'E
18	D4	**Koma**	Oromia	*Jimma*	8°27'N 36°51'E
36	B5	**Kombolcha**	Amhara	*South Wollo*	11°5'N 39°44'E
31	G5	**Kombolcha**	Oromia	*East Hararge*	9°26'N 42°7'E
27	G5	**Kombolcha**	Oromia	*Horo Guduru*	9°30'N 37°28'E
24	E2	**Komosh**	Benishangul Gumuz	*Asosa*	10°18'N 34°38'E
35	H3	**Kon**	Amhara	*North Wollo*	11°36'N 38°56'E
35	I3	**Kon Abo**	Amhara	*North Wollo*	11°45'N 39°9'E
20	B4	**Kondaltiti**	SNNPR	*Gurage*	8°24'N 38°37'E
18	B3	**Kone**	Oromia	*Ilubabor*	8°41'N 36°17'E
41	F5	**Korem**	Tigray	*Southern*	12°32'N 39°30'E
20	E1	**Koremas**	Amhara	*North Shewa (R3)*	9°13'N 39°17'E
26	B3	**Korka**	Benishangul Gumuz	*Metekel*	10°16'N 36°7'E
11	H5	**Kose**	SNNPR	*Gurage*	7°51'N 37°44'E
20	B5	**Koshe**	SNNPR	*Gurage*	8°0'N 38°31'E
4	E1	**Koti**	SNNPR	*Gedio*	6°17'N 38°13'E
18	D3	**Koye**	Oromia	*Jimma*	8°36'N 36°49'E
26	E2	**Kuch**	Amhara	*West Gojam*	10°24'N 37°2'E
21	F5	**Kula**	Oromia	*Arsi*	7°58'N 39°41'E
11	I2	**Kulito**	SNNPR	*Alaba*	7°19'N 38°6'E
35	I2	**Kulmesk**	Amhara	*North Wollo*	11°56'N 39°12'E
30	E5	**Kulubi**	Oromia	*East Hararge*	9°25'N 41°41'E
18	B4	**Kumbabe**	Oromia	*Ilubabor*	8°21'N 36°7'E
19	G4	**Kumbi**	Oromia	*Jimma*	8°7'N 37°29'E
33	J2	**Kunzila**	Amhara	*West Gojam*	11°52'N 37°1'E
28	D1	**Kurkura**	Amhara	*South Wollo*	10°44'N 39°13'E
24	D2	**Kurmuk**	Benishangul Gumuz	*Asosa*	10°33'N 34°17'E
36	A4	**Kuta Ber**	Amhara	*South Wollo*	11°16'N 39°32'E
12	B3	**Kuyera**	Oromia	*West Arsi*	7°14'N 38°39'E
27	I2	**Kuyi**	Amhara	*East Gojam*	10°30'N 37°59'E
41	F2	**Kwiha**	Tigray	*Southern*	13°28'N 39°32'E

L

35	H2	**Lalibela**	Amhara	*North Wollo*	12°2'N 39°2'E
16	B4	**Lare**	Gambella	*Agnuak*	8°20'N 33°57'E
3	H1	**Laska**	SNNPR	*Basketo*	6°18'N 36°36'E
31	J4	**Lefe Isa**	Somali	*Fafan*	9°37'N 42°59'E
22	C5	**Legehida**	Oromia	*Bale*	7°54'N 41°19'E
17	I5	**Leka**	Oromia	*Ilubabor*	8°5'N 35°40'E
12	A4	**Leku**	SNNPR	*Sidama*	6°52'N 38°26'E
20	B3	**Lemen**	Oromia	*South West Shewa*	8°36'N 38°36'E
28	C4	**Lemi**	Amhara	*North Shewa (R3)*	9°48'N 38°54'E
11	I1	**Lera**	SNNPR	*Selti*	7°45'N 37°53'E
28	D2	**Ligwama**	Amhara	*South Wollo*	10°39'N 39°9'E
11	F4	**Loma Bale**	SNNPR	*Dawro*	6°54'N 37°13'E
37	G3	**Loqiya**	Afar	*Awsi (Zone 1)*	11°43'N 40°58'E
27	I3	**Lumame**	Amhara	*East Gojam*	10°15'N 37°56'E

M

38	C4	**Maganan**	Amhara	*North Gonder*	12°46'N 36°24'E
38	C4	**Maganan**	Amhara	*North Gonder*	12°47'N 36°25'E
40	C1	**Mahbere Tsige**	Tigray	*Central*	13°49'N 38°47'E
2	D1	**Maji**	SNNPR	*Bench Maji*	6°10'N 35°35'E
22	D3	**Majo Weldya**	Oromia	*East Hararge*	8°42'N 41°34'E
33	G4	**Mambuk**	Benishangul Gumuz	*Metekel*	11°17'N 36°14'E
33	G5	**Mandura**	Benishangul Gumuz	*Metekel*	11°5'N 36°25'E
27	F1	**Mankusa**	Amhara	*West Gojam*	10°41'N 37°10'E
32	D4	**Mankush**	Benishangul Gumuz	*Metekel*	11°19'N 35°34'E
9	H1	**Masha**	SNNPR	*Sheka*	7°44'N 35°28'E
41	F2	**May Keyah**	Tigray	*Southern*	13°19'N 39°31'E
41	F2	**May Mekdan**	Tigray	*Southern*	13°34'N 39°34'E
39	J2	**May Tsamri**	Tigray	*North Western*	13°34'N 38°11'E
41	F4	**Maychew**	Tigray	*Southern*	12°47'N 39°32'E
11	I3	**Mazoriya**	SNNPR	*Kembata Tembaro*	7°11'N 37°57'E
18	E4	**Mecha**	Oromia	*Jimma*	8°14'N 37°3'E
21	I3	**Mechara**	Oromia	*West Hararge*	8°36'N 40°20'E
11	G2	**Medula**	SNNPR	*Hadiya*	7°19'N 37°30'E
5	G3	**Megado**	Oromia	*Guji*	5°39'N 38°47'E
29	F2	**Mehal Meda**	Amhara	*North Shewa (R3)*	10°19'N 39°39'E
41	F4	**Mehoni**	Tigray	*Southern*	12°48'N 39°39'E
12	C5	**Mejo**	SNNPR	*Sidama*	6°22'N 38°57'E
34	E3	**Mekane Eyesus**	Amhara	*South Gonder*	11°36'N 38°3'E
28	C1	**Mekane Selam**	Amhara	*South Wollo*	10°44'N 38°45'E
40	E2	**MEKELE**	Tigray	*Southern*	13°29'N 39°28'E

17	I1	**Mekenejo**	Oromia	*West Wellega*	9°12'N 35°43'E
20	C4	**Meki**	Oromia	*East Shewa*	8°9'N 38°49'E
18	A3	**Meko**	Oromia	*Ilubabor*	8°41'N 36°1'E
28	A5	**Mekoda**	Oromia	*West Shewa*	9°30'N 38°20'E
29	G1	**Mekoy**	Amhara	*Oromia*	10°42'N 39°47'E
39	G5	**Meksenyit**	Amhara	*North Gonder*	12°23'N 37°33'E
4	B2	**Mele**	SNNPR	*Gamo Gofa*	5°53'N 37°25'E
30	E4	**Melka Jeldu**	Dire Dawa	*Dire Dawa*	9°37'N 41°46'E
4	D3	**Melka Jewe**	SNNPR	*Segen Peoples'*	5°38'N 37°55'E
21	F2	**Melka Jilo**	Amhara	*North Shewa (R3)*	8°53'N 39°36'E
20	B3	**Melka Kunture**	Oromia	*South West Shewa*	8°42'N 38°36'E
31	G5	**Melka Rafu**	Oromia	*East Hararge*	9°28'N 42°11'E
21	H1	**Melka Sede**	Afar	*Gabi (Zone 3)*	9°15'N 40°8'E
21	H1	**Melka Werer**	Afar	*Gabi (Zone 3)*	9°20'N 40°10'E
20	E4	**Melkasa**	Oromia	*East Shewa*	8°24'N 39°20'E
13	G5	**Mena (Dolo Mena)**	Oromia	*Bale*	6°25'N 39°51'E
25	G4	**Mendi**	Oromia	*West Wellega*	9°48'N 35°6'E
28	E4	**Mendida**	Oromia	*North Shewa (R4)*	9°39'N 39°18'E
25	F2	**Menge**	Benishangul Gumuz	*Asosa*	10°21'N 34°45'E
28	D3	**Meranya**	Amhara	*North Shewa (R3)*	10°12'N 39°5'E
12	E2	**Meraro**	Oromia	*Arsi*	7°24'N 39°14'E
34	J4	**Merawi**	Amhara	*West Gojam*	11°24'N 37°9'E
10	E3	**Meri**	SNNPR	*Dawro*	7°2'N 37°6'E
30	B3	**Mersa**	Amhara	*North Wollo*	11°39'N 39°39'E
34	E5	**Mertule Maryam**	Amhara	*East Gojam*	10°51'N 38°16'E
22	C2	**Mesela**	Oromia	*West Hararge*	9°5'N 41°10'E
39	G2	**Mesfinto**	Tigray	*Western*	13°29'N 37°22'E
34	B3	**Meshenti**	Amhara	*West Gojam*	11°28'N 37°16'E
21	F1	**Meteh Bila**	Amhara	*North Shewa (R3)*	9°12'N 39°42'E
21	G2	**Metehara**	Oromia	*East Shewa*	8°54'N 39°54'E
38	B3	**Metema**	Amhara	*North Gonder*	12°56'N 36°10'E
10	D2	**Metoso**	Oromia	*Jimma*	7°25'N 36°52'E
17	I4	**Metu (Mettu)**	Oromia	*Ilubabor*	8°18'N 35°35'E
29	G3	**Mezezo**	Amhara	*North Shewa (R3)*	10°0'N 39°44'E
21	I3	**Micheta**	Oromia	*Arsi*	8°33'N 40°18'E
28	D3	**Mida**	Amhara	*North Shewa (R3)*	10°18'N 39°8'E
22	A1	**Mieso**	Oromia	*West Hararge*	9°13'N 40°45'E
37	F4	**Mile**	Afar	*Awsi (Zone 1)*	11°25'N 40°45'E
27	J5	**Minare**	Oromia	*West Shewa*	9°36'N 38°16'E
9	I3	**Mizan Teferi**	SNNPR	*Bench Maji*	7°00'N 35°25'E
20	D3	**Modjo**	Oromia	*East Shewa*	8°35'N 39°7'E
29	F3	**Molale**	Amhara	*North Shewa (R3)*	10°7'N 39°40'E
11	F5	**Morka**	SNNPR	*Gamo Gofa*	6°25'N 37°18'E
11	H1	**Morsito**	SNNPR	*Hadiya*	7°42'N 37°49'E
34	D5	**Mota**	Amhara	*East Gojam*	11°4'N 37°52'E
18	E2	**Mote**	Oromia	*East Wellega*	8°49'N 36°54'E
5	J4	**Mugayo**	Oromia	*Guji*	5°12'N 39°25'E
28	B5	**Muger**	Oromia	*West Shewa*	9°28'N 38°36'E
16	E3	**Mugi**	Oromia	*Kelem Wellega*	8°34'N 34°36'E
35	I2	**Muja**	Amhara	*North Wollo*	12°0'N 39°17'E
28	C5	**Muke Turi**	Oromia	*North Shewa (R4)*	9°33'N 38°52'E
17	I1	**Muklemi**	Oromia	*West Wellega*	9°17'N 35°32'E
28	B5	**Mulo**	Oromia	*North Shewa (R4)*	9°22'N 38°39'E
12	C2	**Munesa**	Oromia	*Arsi*	7°34'N 38°54'E
N					
11	F2	**Nada**	Oromia	*Jimma*	7°37'N 37°14'E
19	G5	**Natri**	Oromia	*Jimma*	8°2'N 37°27'E
35	F3	**Nefas Mewcha**	Amhara	*South Gonder*	11°43'N 38°28'E
6	A4	**Negele**	Oromia	*Guji*	5°20'N 39°34'E
25	H5	**Nejo**	Oromia	*West Wellega*	9°30'N 35°30'E
18	C2	**Nekemte**	Oromia	*East Wellega*	9°5'N 36°34'E
17	I4	**Nopa**	Oromia	*Ilubabor*	8°23'N 35°36'E
24	E5	**Nopa**	Oromia	*West Wellega*	9°30'N 34°38'E
18	D3	**Nunu**	Oromia	*East Wellega*	8°45'N 36°38'E
O					
26	E3	**Obera**	Oromia	*Horo Guduru*	10°0'N 37°1'E
20	D5	**Ogokho**	Oromia	*Arsi*	8°3'N 38°59'E
20	C4	**Ombole**	Oromia	*East Shewa*	8°22'N 38°46'E
32	C2	**Omedila**	Amhara	*North Gonder*	11°55'N 35°16'E
3	F5	**Omorate**	SNNPR	*South Omo*	4°48'N 36°3'E
3	J1	**Otolo**	SNNPR	*Gamo Gofa*	6°10'N 37°6'E

P					
33	G4	**Pawe**	Benishangul Gumuz	*Metekel*	11°14'N 36°16'E
R					
29	F2	**Rabel**	Amhara	*North Shewa (R3)*	10°29'N 39°39'E
28	C3	**Rema**	Amhara	*North Shewa (R3)*	10°15'N 38°54'E
13	F4	**Rira**	Oromia	*Bale*	6°46'N 39°44'E
19	H2	**Rob Geba**	Oromia	*West Shewa*	9°7'N 37°38'E
27	H2	**Rob Gebeya**	Amhara	*East Gojam*	10°32'N 37°45'E
18	E1	**Rob Gebeya**	Oromia	*East Wellega*	9°9'N 36°55'E
17	F3	**Rob Gebeya**	Oromia	*Kelem Wellega*	8°41'N 34°52'E
13	H3	**Robe**	Oromia	*Bale*	7°7'N 40°0'E
21	F5	**Robi**	Oromia	*Arsi*	7°51'N 39°37'E
34	B1	**Robit**	Amhara	*North Gonder*	12°18'N 37°23'E
36	A2	**Robit**	Amhara	*North Wollo*	12°0'N 39°37'E
11	I3	**Ropi**	Oromia	*West Arsi*	7°9'N 38°6'E
S					
12	D1	**Sagure**	Oromia	*Arsi*	7°44'N 39°9'E
19	G5	**Saja**	Oromia	*Jimma*	7°58'N 37°26'E
40	D3	**Samre**	Tigray	*Southern*	13°11'N 39°12'E
39	I3	**Sankaber**	Amhara	*North Gonder*	13°13'N 38°2'E
29	F4	**Sasit**	Amhara	*North Shewa (R3)*	9°55'N 39°29'E
3	I1	**Sawla**	SNNPR	*Gamo Gofa*	6°17'N 36°52'E
20	B2	**Sebeta**	Oromia	*South West Shewa*	8°54'N 38°36'E
34	D5	**Sede**	Amhara	*East Gojam*	10°55'N 37°53'E
13	G2	**Sede**	Oromia	*Arsi*	7°32'N 39°58'E
16	E3	**Sedi (Ashi)**	Oromia	*Kelem Wellega*	8°39'N 34°40'E
13	G2	**Sedika**	Oromia	*Arsi*	7°37'N 39°48'E
15	I1	**Segeg**	Somali	*Nogob*	7°41'N 42°50'E
4	B3	**Segen**	SNNPR	*Segen Peoples'*	5°35'N 37°32'E
35	J3	**Segora**	Amhara	*South Wollo*	11°34'N 39°20'E
10	D2	**Seka**	Oromia	*Jimma*	7°36'N 36°43'E
18	E4	**Seka**	Oromia	*Jimma*	8°12'N 36°56'E
19	G5	**Sekoru**	Oromia	*Jimma*	7°55'N 37°25'E
40	D4	**Sekota**	Amhara	*Wag Himra*	12°37'N 39°1'E
29	F4	**Sela Dingay**	Amhara	*North Shewa (R3)*	9°57'N 39°37'E
11	G5	**Selam Ber**	SNNPR	*Gamo Gofa*	6°28'N 37°27'E
13	H3	**Selka**	Oromia	*Bale*	7°2'N 40°12'E
10	D2	**Sembo**	Oromia	*Jimma*	7°34'N 36°37'E
37	G3	**SEMERA**	Afar	*Awsi (Zone 1)*	11°47'N 41°0'E
29	G3	**Senbete**	Amhara	*Oromia*	10°18'N 39°59'E
20	D2	**Sendafa**	Oromia	*North Shewa (R4)*	9°9'N 39°1'E
41	F1	**Senkata (Negash)**	Tigray	*Eastern*	13°53'N 39°36'E
27	H2	**Senyin**	Amhara	*East Gojam*	10°35'N 37°41'E
10	D1	**Senyo Gebeya**	Oromia	*Jimma*	7°43'N 36°37'E
38	E5	**Seraba**	Amhara	*North Gonder*	12°33'N 37°04'E
10	E1	**Serbo**	Oromia	*Jimma*	7°42'N 36°59'E
13	H1	**Seru**	Oromia	*Arsi*	7°40'N 40°12'E
18	B5	**Setema**	Oromia	*Jimma*	8°2'N 36°20'E
19	F3	**Seyo**	Oromia	*West Shewa*	8°46'N 37°15'E
5	H3	**Shakiso**	Oromia	*Guji*	5°46'N 38°53'E
26	E5	**Shambu**	Oromia	*Horo Guduru*	9°33'N 37°5'E
12	B3	**Shashemene**	Oromia	*West Arsi*	7°11'N 38°35'E
10	C2	**Shebe**	Oromia	*Jimma*	7°31'N 36°31'E
16	E3	**Shebel**	Oromia	*Kelem Wellega*	8°28'N 34°35'E
19	F2	**Sheboka**	Oromia	*West Shewa*	9°5'N 37°7'E
41	F2	**Shehet**	Afar	*Kilbati (Zone 2)*	13°21'N 39°45'E
14	A1	**Sheki Husen**	Oromia	*Bale*	7°44'N 40°42'E
27	G2	**Shelel**	Amhara	*East Gojam*	10°34'N 37°35'E
19	G3	**Shenen**	Oromia	*West Shewa*	8°44'N 37°27'E
20	E5	**Sheno**	Oromia	*North Shewa (R4)*	9°19'N 39°17'E
9	I3	**Shewa Gimira**	SNNPR	*Bench Maji*	6°57'N 35°46'E
29	G3	**Shewa Robit**	Amhara	*North Shewa (R3)*	10°0'N 39°54'E
27	I5	**Shikute**	Oromia	*West Shewa*	9°22'N 38°1'E
28	D2	**Shil Afaf**	Amhara	*South Wollo*	10°22'N 39°10'E
26	E1	**Shindi**	Amhara	*West Gojam*	10°39'N 36°57'E
31	F4	**Shinile**	Somali	*Siti*	9°49'N 41°52'E
20	A1	**Shino**	Oromia	*West Shewa*	9°19'N 38°16'E
12	D3	**Shire**	Oromia	*West Arsi*	7°14'N 38°59'E
20	E1	**Shola Gebeya**	Amhara	*North Shewa (R3)*	9°12'N 39°24'E
11	I3	**Shone**	SNNPR	*Hadiya*	7°8'N 37°57'E
17	G4	**Sibo**	Oromia	*Ilubabor*	8°14'N 35°14'E

17	G2	**Sichawo**	Oromia	*Kelem Wellega*	9°6'N 35°1'E
18	A5	**Sigamo**	Oromia	*Jimma*	7°54'N 36°5'E
19	G3	**Silk Amba**	Oromia	*West Shewa*	8°32'N 37°25'E
12	E2	**Siltana**	Oromia	*Arsi*	7°23'N 39°23'E
18	C4	**Sineso**	Oromia	*Ilubabor*	8°8'N 36°33'E
20	E4	**Sire**	Oromia	*Arsi*	8°16'N 39°29'E
18	D2	**Sire**	Oromia	*East Wellega*	9°2'N 36°52'E
36	A3	**Sirinka**	Amhara	*North Wollo*	11°45'N 39°36'E
40	C4	**Siska**	Amhara	*Wag Himra*	12°45'N 38°48'E
	G3	**Soda**	Oromia	*Borena*	5°29'N 38°36'E
11	H4	**Soddo**	SNNPR	*Wolayita*	6°51'N 37°45'E
20	E4	**Sodere**	Oromia	*East Shewa*	8°24'N 39°22'E
5	G2	**Solemo**	Oromia	*Guji*	6°6'N 38°35'E
26	E4	**Sombo**	Oromia	*Horo Guduru*	9°44'N 37°1'E
10	D3	**Soyema**	SNNPR	*Konta*	7°16'N 36°49'E
	D3	**Soyema**	SNNPR	*Segen Peoples'*	5°29'N 37°51'E
36	B4	**Sulula**	Amhara	*South Wollo*	11°14'N 39°41'E
20	C1	**Sululta**	Oromia	*North Shewa (R4)*	9°10'N 38°45'E
17	I3	**Supe**	Oromia	*Ilubabor*	8°31'N 35°39'E
26	B4	**Surge**	Oromia	*East Wellega*	9°56'N 36°10'E
	E5	**Surupa**	Oromia	*Borena*	5°8'N 38°18'E

T

27	H1	**Tame**	Amhara	*West Gojam*	10°46'N 37°38'E
29	F4	**Tarmaber**	Amhara	*North Shewa (R3)*	9°49'N 39°43'E
19	H4	**Tawula**	SNNPR	*Gurage*	8°16'N 37°41'E
1	H4	**Tebela**	SNNPR	*Wolayita*	6°42'N 37°46'E
19	G4	**Tedele**	SNNPR	*Gurage*	8°18'N 37°37'E
	E5	**Tedim**	Oromia	*Borena*	5°1'N 38°9'E
2	A5	**Teferi Kela**	SNNPR	*Sidama*	6°29'N 38°23'E
20	B2	**Tefki**	Oromia	*South West Shewa*	8°50'N 38°29'E
	B5	**Teltele**	Oromia	*Borena*	5°3'N 37°22'E
	I3	**Temenja Yazhi**	SNNPR	*Bench Maji*	7°4'N 35°38'E
37	G3	**Tendaho**	Afar	*Awsi (Zone 1)*	11°41'N 40°56'E
5	I4	**Tenta**	Amhara	*South Wollo*	11°19'N 39°15'E
	H2	**Tepi**	Gambella	*Majang*	7°12'N 35°25'E
2	E1	**Tewodros Ketema**	Amhara	*North Gonder*	12°13'N 35°53'E
7	F2	**Tibe**	Oromia	*Kelem Wellega*	8°58'N 34°45'E
9	F2	**Tibe**	Oromia	*West Shewa*	9°4'N 37°11'E
3	F1	**Ticho**	Oromia	*Arsi*	7°48'N 39°30'E
9	F3	**Tikil Dingay**	Amhara	*North Gonder*	12°44'N 37°24'E
3	J5	**Tilili**	Amhara	*Awi/Agew*	10°51'N 37°1'E
1	I4	**Timuga**	Oromia	*Arsi*	8°25'N 40°14'E
9	F5	**Tiro**	Oromia	*Jimma*	8°4'N 37°12'E
4	C3	**Tis Abay**	Amhara	*West Gojam*	11°29'N 37°34'E
0	B4	**Tiya**	SNNPR	*Gurage*	8°26'N 38°36'E
8	C5	**Toba**	Oromia	*Jimma*	8°0'N 36°29'E
5	G4	**Toba**	Oromia	*West Wellega*	9°41'N 35°15'E
1	G1	**Toba**	SNNPR	*Yem*	7°44'N 37°28'E
6	E1	**Tobi**	Oromia	*West Wellega*	9°15'N 34°34'E
0	E3	**Tocha**	SNNPR	*Dawro*	7°5'N 37°1'E
1	F3	**Tocha Sodo**	SNNPR	*Dawro*	7°8'N 37°9'E
9	F4	**Toley**	Oromia	*Jimma*	8°17'N 37°22'E
	H5	**Toom**	SNNPR	*Bench Maji*	6°18'N 35°24'E
2	A1	**Tora**	SNNPR	*Selti*	7°51'N 38°25'E
	E2	**Tore**	Oromia	*Borena*	5°55'N 38°9'E
2	B4	**Tula**	SNNPR	*Sidama*	6°57'N 38°28'E
8	B3	**Tulema**	Oromia	*Ilubabor*	8°45'N 36°11'E
1	F3	**Tulema**	SNNPR	*Dawro*	7°0'N 37°14'E
9	J3	**Tulu Bolo**	Oromia	*South West Shewa*	8°39'N 38°12'E
8	J4	**Tulu Miki**	Oromia	*North Shewa (R4)*	9°54'N 38°20'E
	D1	**Tum**	SNNPR	*Bench Maji*	6°15'N 35°31'E
	H5	**Turmi**	SNNPR	*South Omo*	4°48'N 36°29'E

U

7	H4	**Uka**	Oromia	*Ilubabor*	8°10'N 35°21'E

W

	B1	**Wacha**	SNNPR	*Gamo Gofa*	6°19'N 37°18'E
	I3	**Wacha**	SNNPR	*Kaffa*	7°7'N 35°47'E
	I3	**Wadera**	Oromia	*Guji*	5°46'N 39°18'E
6	A1	**Waja**	Tigray	*Southern*	12°17'N 39°36'E
11	H5	**Wajifo**	SNNPR	*Gamo Gofa*	6°27'N 37°44'E
11	F3	**Waka**	SNNPR	*Dawro*	7°3'N 37°10'E
18	D3	**Wama Adere**	Oromia	*East Wellega*	8°45'N 36°45'E
34	C3	**Wanzaye**	Amhara	*South Gonder*	11°46'N 37°41'E
30	E5	**Water**	Oromia	*East Hararge*	9°22'N 41°46'E
28	D4	**Weberi**	Oromia	*North Shewa (R4)*	9°40'N 39°1'E
27	I2	**Webrje**	Amhara	*East Gojam*	10°23'N 38°3'E
35	I3	**Wegel Tena**	Amhara	*North Wollo*	11°35'N 39°13'E
28	C2	**Wegeti**	Amhara	*South Wollo*	10°35'N 38°45'E
38	D4	**Wehini**	Amhara	*North Gonder*	12°37'N 36°39'E
39	H3	**Wekin**	Amhara	*North Gonder*	13°3'N 37°49'E
11	F3	**Welde Hane**	SNNPR	*Dawro*	7°14'N 37°16'E
36	A2	**Weldiya**	Amhara	*North Wollo*	11°49'N 39°36'E
20	E3	**Welenchiti**	Oromia	*East Shewa*	8°39'N 39°25'E
28	B5	**Welenkombi**	Oromia	*West Shewa*	9°24'N 38°29'E
19	I3	**Weliso**	Oromia	*South West Shewa*	8°32'N 37°58'E
19	H4	**Welkite**	SNNPR	*Gurage*	8°16'N 37°46'E
4	E1	**Wenago**	SNNPR	*Gedio*	6°18'N 38°15'E
12	B3	**Wendo Genet**	SNNPR	*Sidama*	7°2'N 38°36'E
20	D4	**Wenji**	Oromia	*East Shewa*	8°27'N 39°13'E
20	D1	**Wenoda**	Oromia	*North Shewa (R4)*	9°17'N 39°5'E
28	E2	**Were Ilu**	Amhara	*South Wollo*	10°35'N 39°26'E
28	B2	**Werk Amba**	Amhara	*South Wollo*	10°38'N 38°34'E
40	D1	**Werk Amba**	Tigray	*Central*	13°44'N 39°00'E
12	D4	**Werka**	Oromia	*West Arsi*	6°43'N 39°10'E
33	J4	**Wetet Abay**	Amhara	*West Gojam*	11°22'N 37°2'E
41	G5	**Weyra Wuha**	Tigray	*Southern*	12°32'N 39°46'E
3	H2	**Wib Hamer**	SNNPR	*South Omo*	6°3'N 36°35'E
36	A3	**Wichale**	Amhara	*South Wollo*	11°31'N 39°36'E
12	B4	**Wijigra**	SNNPR	*Sidama*	6°56'N 38°33'E
36	A3	**Wirgesa**	Amhara	*North Wollo*	11°32'N 39°37'E
10	E4	**Wishaye Bele**	SNNPR	*Dawro*	6°54'N 37°0'E
3	J4	**Woito**	SNNPR	*South Omo*	5°22'N 36°59'E
19	J2	**Wolenkomi**	Oromia	*West Shewa*	9°00'N 38°15'E
20	E5	**Wolonkomi**	Oromia	*Arsi*	8°2'N 39°15'E
11	J1	**Worabe**	SNNPR	*Selti*	7°51'N 38°11'E
34	C2	**Woreta**	Amhara	*South Gonder*	11°55'N 37°41'E
41	F1	**Wukro**	Tigray	*Eastern*	13°46'N 39°35'E
11	I1	**Wulaberg**	SNNPR	*Selti*	7°44'N 38°07'E
10	B2	**Wushwush**	SNNPR	*Kaffa*	7°18'N 36°8'E

Y

4	E5	**Yabelo**	Oromia	*Borena*	4°53'N 38°5'E
18	C4	**Yanfa**	Oromia	*Ilubabor*	8°14'N 36°35'E
17	H2	**Yara**	Oromia	*West Wellega*	9°5'N 35°23'E
17	J4	**Yayu**	Oromia	*Ilubabor*	8°20'N 35°49'E
27	H2	**Yebokile**	Amhara	*East Gojam*	10°25'N 37°51'E
10	D1	**Yebu**	Oromia	*Jimma*	7°47'N 36°43'E
40	D3	**Yechilay**	Tigray	*Central*	13°17'N 38°59'E
27	H3	**Yejube**	Amhara	*East Gojam*	10°8'N 37°44'E
27	J3	**Yekoza**	Amhara	*East Gojam*	10°14'N 38°12'E
18	C4	**Yembero**	Oromia	*Ilubabor*	8°10'N 36°27'E
27	J2	**Yetmen**	Amhara	*East Gojam*	10°20'N 38°8'E
27	G2	**Yewla**	Amhara	*East Gojam*	10°24'N 37°34'E
34	C2	**Yifag**	Amhara	*South Gonder*	12°4'N 37°44'E
5	G1	**Yirba Moda**	Oromia	*Guji*	6°12'N 38°42'E
12	A4	**Yirga Alem**	SNNPR	*Sidama*	6°44'N 38°24'E
4	E1	**Yirgacheffe**	SNNPR	*Gedio*	6°9'N 38°12'E
33	I3	**Yismala**	Amhara	*West Gojam*	11°36'N 36°54'E
17	H2	**Yubdo**	Oromia	*West Wellega*	8°57'N 35°26'E

Z

39	I2	**Zarima**	Amhara	*North Gonder*	13°20'N 37°53'E
11	H5	**Zefine**	SNNPR	*Gamo Gofa*	6°28'N 37°38'E
34	B3	**Zege (Zeghie)**	Amhara	*West Gojam*	11°41'N 37°19'E
5	I2	**Zembaba Wiha**	Oromia	*Guji*	5°53'N 39°10'E
26	B1	**Zigin**	Benishangul Gumuz	*Metekel*	10°40'N 36°9'E
20	B5	**Ziway**	Oromia	*East Shewa*	7°55'N 38°42'E

The Coffee Landscape of Ethiopia

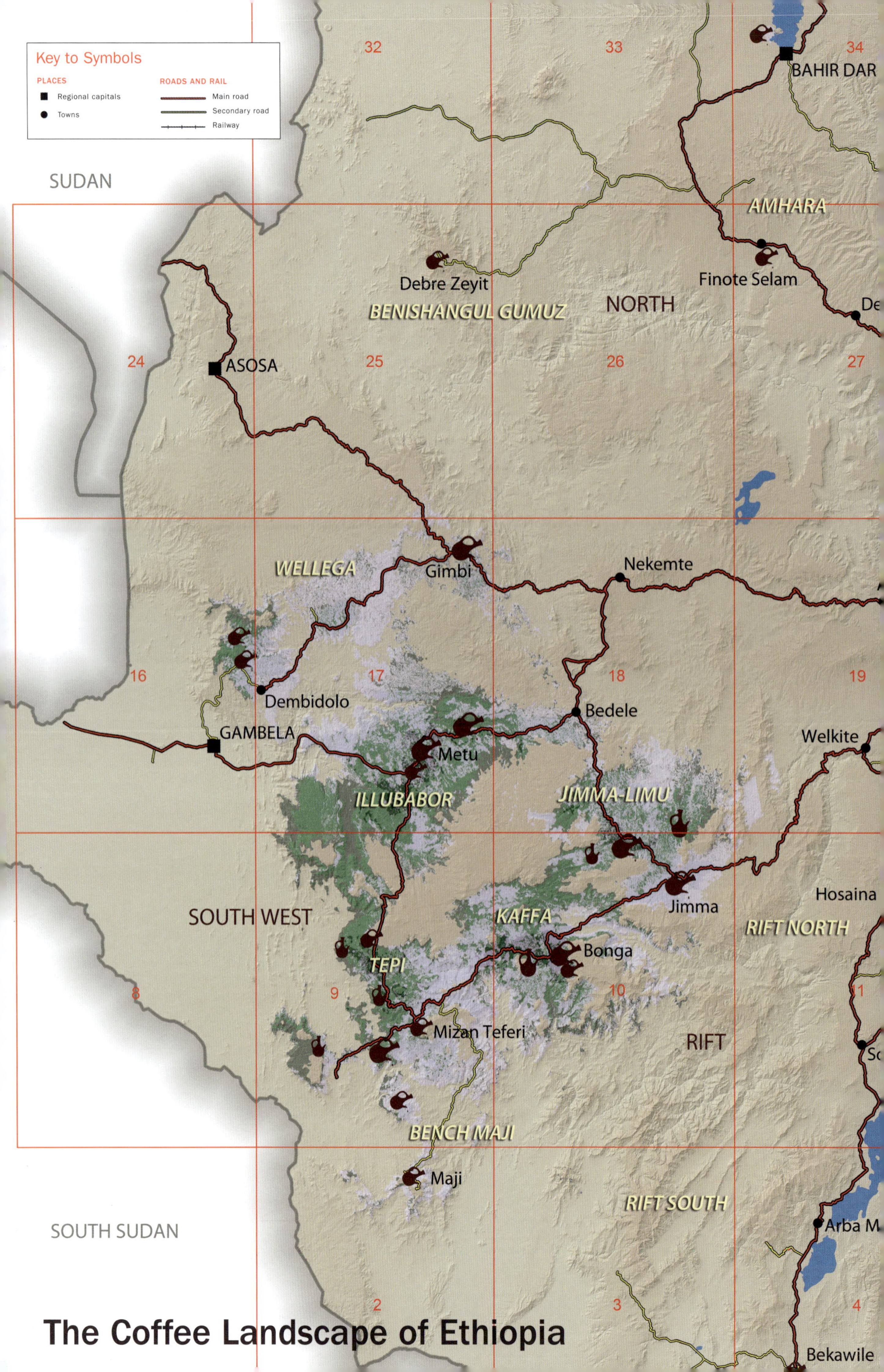